The Laboratory
MOUSE

T0121404

With the advent of transgenic and other genetic engineering technologies, the versatility and usefulness of the mouse as a model in biomedical research has soared. Revised to reflect advances since the second edition, *The Laboratory Mouse* continues to be the most accessible reference on the biology and care of the mouse in research settings.

This guide presents basic information and common procedures in detail to provide a quick reference source for investigators, technicians, and caretakers on the humane care and use of the mouse. The new edition adds information on novel technologies such as CRISPR-Cas and on housing systems and management practices; it covers new concepts such as pain assessment by facial expression and the importance of nest building as an assessment tool of well-being. There are now expanded sections on anesthesia and analgesia, and on behavior and enrichment.

An ideal quick reference for investigators, technicians, and animal caretakers charged with the care and/or use of mice in a research setting, this book will be particularly valuable to those new to working with mice who need to start research programs using these animals.

Laboratory Animal Pocket Reference Series

Series Editor: **Mark A. Suckow**

PUBLISHED TITLES

The Laboratory Rabbit, Second Edition,
By Mark A. Suckow, Valerie Schroeder, Fred A. Douglas

The Laboratory Zebrafish,
By Claudia Harper, Christian Lawrence

The Laboratory Guinea Pig, Second Edition,
By Donna J. Clemons, Jennifer L. Seeman

The Laboratory Ferret,
*By C. Andrew Matchett, Rena Marr, Felipe M. Berard,
Andrew G. Cawthon, Sonya P. Swing*

The Laboratory Mouse, Second Edition,
By Peggy J. Danneman, Mark A. Suckow, Cory Brayton

Critical Care Management for Laboratory Mice and Rats,
By F. Claire Hankenson

The Laboratory Bird,
By Douglas K. Taylor, Vanessa K. Lee, Karen R. Strait

**Handbook of Laboratory Animal Anesthesia
and Pain Management: Rodents,**
Edited By Cholawat Pacharinsak, Jennifer C. Smith

The Laboratory Nonhuman Primate, Second Edition,
By Jeffrey D. Fortman, Terry A. Hewett, Lisa C. Halliday

Laboratory Animals Pocket Reference Set,
By Various Authors

The Laboratory Mouse, Third Edition,
By Mark A. Suckow, Sara Hashway, Kathleen R. Pritchett-Corning

For more details, please visit: Laboratory Animal Pocket
Reference—Book Series—Routledge & CRC Press.

The Laboratory
MOUSE

[Third Edition]

Mark A. Suckow,
Sara A. Hashway, and
Kathleen R. Pritchett-Corning

CRC Press
Taylor & Francis Group
Boca Raton London New York

CRC Press is an imprint of the
Taylor & Francis Group, an **informa** business

Third edition published 2023
by CRC Press
6000 Broken Sound Parkway NW, Suite 300, Boca Raton, FL 33487-2742

and by CRC Press
4 Park Square, Milton Park, Abingdon, Oxon, OX14 4RN

CRC Press is an imprint of Taylor & Francis Group, LLC

ISBN: 978-1-032-41685-4 (hbk)
ISBN: 978-0-367-37320-7 (pbk)
ISBN: 978-0-429-35308-6 (ebk)

DOI: 10.1201/9780429353086

Typeset in Bookman
by Deanta Global Publishing Services, Chennai, India

contents

preface...xiii
about the authors xv

1 important biological features 1

    introduction 1
    stocks and strains...2
    nomenclature 3
    anatomic and physiologic features 5
    normative values 9
        Basic Biologic Parameters 9
        Clinical Chemistry....................................9
        Urinalysis 11
    hematology...12
    immunodeficient mice.......................................13
    wild mice...15
    behavior, well-being, and enrichment 17
        Behavior 17
        Well-being..19
        Enrichment...20
    references 21

2 husbandry..25

    housing 25
        Caging..25
        Housing Systems 28

Bedding 30
barriers and containment 33
personal protective equipment (PPE) 36
special considerations for immunodeficient mice 37
environment 38
Temperature and Humidity 38
Ventilation 39
Illumination 39
Noise and Vibration 40
sanitation and pest control 41
Sanitation 41
Cage cleaning 41
Room cleaning 43
Research equipment cleaning 43
Quality control 43
Pest Control 44
nutrition 44
water 47
breeding 47
Basic Genetics 48
Breeding Schemes 49
Pheromone Influences 52
Timed Pregnancy 53
Genetic Monitoring 54
identification and record keeping 55
Identification 55
Records 57
transportation 58
Between Institutions 58
Within the Institution 60
references 60

3   management 69

regulatory agencies and compliance 69
Institutional Animal Care and Use Committee (IACUC) 71
IACUC Composition 71
Responsibilities of the IACUC 72
occupational health and zoonotic diseases 72
compassion fatigue 75
references 76

4 clinical medicine 79

basic veterinary supplies 79
physical examination of the mouse.....................................80
common spontaneous and noninfectious diseases...............82
   Common Conditions Involving the Alimentary System 84
      Non-neoplastic conditions..84
      Alimentary tumors...85
   Common Conditions Involving the Cardiovascular
   System...85
      Non-neoplastic conditions..85
      Cardiovascular tumors...87
   Common Conditions Involving the Endocrine System......87
      Non-neoplastic conditions..87
      Endocrine tumors 87
   Hematopoietic and Immune System 88
      Non-neoplastic conditions..88
      Hematopoietic and immune system tumors...............89
   Common Conditions Involving the Integumentary
   System...89
      Non-neoplastic conditions..89
      Tumors of skin and adnexae.....................................91
   Common Conditions Involving the Musculoskeletal
   System...93
      Non-neoplastic conditions..93
      Musculoskeletal tumors 94
   Common Conditions Involving the Nervous System.........94
      Non-neoplastic conditions..94
      Nervous system tumors..96
   Common Conditions Involving the Respiratory System....97
      Non-neoplastic conditions..97
      Respiratory tumors 97
   Common Conditions Involving the Urogenital System 97
      Non-neoplastic conditions..97
      Urogenital tumors 98
   Common Systemic or Multisystem Conditions................98
      Amyloidosis...98
      Hyalinosis 99
      Conditions related to nutritional status....................99
      Stress-related changes 100
common infectious diseases 100

Diagnostic Methods 102
Viral Agents and Diseases 103
   Murine noroviruses (MNV) 104
   Parvoviruses 104
   Mouse hepatitis virus (MHV) 105
   Theiler's mouse encephalomyelitis virus (TMEV) 106
   Mouse rotavirus (MRV, EDIM) 106
   Murine astrovirus 107
   Mouse adenoviruses (MAV 1 and 2) 107
   Lymphocytic choriomeningitis virus (LCMV) 108
   Lactate dehydrogenase–elevating virus (LDV, LDEV) 108
   Hantaviruses (Han) 108
   Herpesviruses 108
   Ectromelia virus (mousepox) 108
   Papovaviruses 109
   Paramyxoviruses 110
   Reovirus 3 (REO3)... 110
Bacterial Agents and Diseases 110
   Bordetella species 110
   Filobacterium rodentium (formerly cilia-associated respiratory bacillus, CAR bacillus, CARB) 111
   Citrobacter rodentium... 111
   Clostridium piliforme 112
   Corynebacterium bovis.. 112
   Helicobacter... 112
   Mycoplasma 113
   Rodentibacter pneumotropicus and *R. heylii* (formerly *Pasteurella pneumotropica* type Jawetz or Heyl)... 113
   Pseudomonas aeruginosa....................................... 114
   Staphylococcus 114
Fungal Agents and Diseases 115
   Pneumocystis murina.. 115
   Intestinal protozoa 115
Parasitic Agents and Diseases 117
   Arthropods.. 117
   Helminths 119
treatment and supportive care of sick mice...................... 122
Drug Dosages 122
General Treatment of Open Skin Lesions 122

General Treatment of Weak Mice.................................... 122
    Provision of supplemental fluids 123
    Provision of food... 123
    Provision of supplemental heat 124
Treatment of Dystocia.. 124
clinical endpoints 125
treating disease on a colony basis.................................. 129
Prevention of Spread 129
Approaches to Elimination... 131
    Depopulation.. 131
    Test and cull 131
    Burnout 132
    Medical treatment 132
Facility Decontamination... 133
references... 133

5   preventive medicine.. 141

receiving.. 141
Reviewing Health Reports 141
Options for Newly Imported Mice.................................. 144
    Transport directly to barrier 144
    Quarantine 144
    Rederivation... 146
    Cryopreservation... 148
testing of biological materials 148
Risks for Humans and Animals 148
health surveillance and monitoring.................................. 150
Principles.. 150
Selection of Test Subjects... 151
Diagnostic Testing 153
disease prevention through sanitation.............................. 153
references... 154

6   experimental methodology... 157

restraint.. 157
Cage Transfer 157
Restraint for Manipulation.. 158
Restraint Devices.. 159
sampling methods 159
Blood 159

Urine 165
Feces 167
Samples for DNA Analysis... 167
Vaginal Swabs 168
compound administration 169
Oral (PO)... 169
Intramuscular (IM) 170
Intraperitoneal (IP).. 171
Subcutaneous (SC) 172
Intradermal (ID)... 172
Intravenous (IV) 172
Retro-orbital (RO).. 173
Implantable Cannulas and Pumps................................ 174
anesthesia and analgesia 175
Anesthesia... 175
Methods and drugs 176
Periprocedural Care.. 182
Analgesia.. 185
euthanasia 188
Laws, Regulations, and Guidelines............................... 189
Management Considerations... 189
Scientific Considerations.. 190
Experimental Endpoints 190
Euthanasia Methods.. 191
necropsy... 193
Equipment and Materials.. 193
Necropsy Procedure 195
External examination 196
Dissection and specimen collection 196
Trimming tissues for histology 199
Histopathology 200
Reporting and archiving data and specimens..........200
references...200

7   resources and additional information 209

organizations..209
American Association for Laboratory Animal Science
(AALAS), www.aalas.org...209
Laboratory Animal Management Association (LAMA),
www.lama-online.org 210

American Society of Laboratory Animal Practitioners
(ASLAP), www.aslap.org 210
American College of Laboratory Animal Medicine
(ACLAM), www.aclam.org 210
Laboratory Animal Welfare Training Exchange
(LAWTE), www.lawte.org 211
Institute for Laboratory Animal Research (ILAR),
www.nationalacademies.org/ilar/institute-for
-laboratory-animal-research 211
AAALAC International, www.aaalac.org 211
Foundation for Biomedical Research (FBR),
fbresearch.org 212
National Association for Biomedical Research (NABR),
www.nabr.org 212
publications 212
Books 212
Guideline Documents 214
Periodicals 214
electronic resources 215

appendix 217
suggested cassette numbering system and some
trimming suggestions 217
index 221

preface

The use of laboratory animals, including mice, continues to be an important part of biomedical research. In many instances, individuals performing such research are charged with broad responsibilities, including animal facility management, animal husbandry, regulatory compliance, and performance of technical procedures directly related to research projects. In this regard, this handbook was written to provide a quick reference for investigators, technicians, and animal caretakers charged with the care and/or use of mice in a research setting. It should be particularly valuable to those at small institutions or facilities lacking a large, well-organized animal resource unit and to those individuals who need to conduct research programs using mice and are starting from scratch.

This handbook is organized into seven chapters: "important biological features" (Chapter 1), "husbandry" (Chapter 2), "management" (Chapter 3), "clinical medicine" (Chapter 4), "preventive medicine" (Chapter 5), "experimental methodology" (Chapter 6), and "resources and additional information" (Chapter 7), which includes organizations, periodicals, books, and online resources.

A final point to be considered is that all individuals performing procedures described in this handbook should be properly trained. Mouse welfare is improved by initial and continuing education of personnel, and this will facilitate the overall success of programs using mice in research, teaching, or testing.

The authors wish to specifically acknowledge Drs. Peggy Danneman and Cory Brayton who co-authored the first and second editions of this book along with Dr. Mark Suckow. The third edition builds upon the work of these earlier authors, and it is our hope that this edition will serve as a valuable reference for those charged with the humane care and use of mice in research, teaching, and testing.

about the authors

Mark A. Suckow, D.V.M., is Associate Vice President for Research, Attending Veterinarian, and Professor of Biomedical Engineering at the University of Kentucky in Lexington, KY. Dr. Suckow earned the degree of Doctor of Veterinary Medicine from the University of Wisconsin in Madison, WI, in 1987 and completed a post-doctoral residency program in laboratory animal medicine at the University of Michigan in Ann Arbor, MI, in 1990. He is a diplomate of the American College of Laboratory Animal Medicine. Dr. Suckow has published over 100 scientific papers and chapters in books. He is a past president of the American Association for Laboratory Animal Science and of the American Society of Laboratory Animal Practitioners and serves on the Council on Accreditation of AAALAC International.

Sara A. Hashway, D.V.M., is Director of the Office of Animal Resources, Attending Veterinarian, and Assistant Research Professor of Psychology and Neuroscience at the University of Colorado Boulder in Boulder, CO. Dr. Hashway earned her veterinary degree from the University of Georgia in Athens, GA, in 2010 and completed a post-doctoral fellowship in laboratory animal medicine at the University of Michigan in Ann Arbor, MI, in 2013. She is a diplomate of the American College of Laboratory Animal Medicine and serves as an ad hoc consultant for AAALAC International. Dr. Hashway has published multiple scientific papers and co-authored the chapter "The Translational Potential of Rats" for the third edition of the ACLAM Series book *The Laboratory Rat*.

Kathleen R. Pritchett-Corning, D.V.M., is Attending Veterinarian and Director, Office of Animal Resources, at the Harvard University Faculty of Arts and Sciences, in Cambridge, MA, and Affiliate Assistant Professor in the Department of Comparative Medicine at the University of Washington in Seattle, WA. Dr. Pritchett-Corning received her B.S. and her D.V.M. from Washington State University in Pullman, WA, and her post-doctoral training from the University of Washington. She became a diplomate of the American College of Laboratory Animal Medicine in 2002 and has held positions at the University of Washington, in Seattle, WA, the Jackson Laboratory in Bar Harbor, ME, and Charles River Laboratories, in Wilmington, MA. In 2015, she received the AALAS Pravin Bhatt Scientific Excellence Award. Dr. Pritchett-Corning has worked primarily with mice for 30 years and has authored more than 80 peer-reviewed publications.

important biological features

introduction

Mice have been domesticated for centuries, perhaps even for millennia, and have been used in scientific research since the 1600s (Keeler 1931). However, development of the laboratory mouse as a research model really began with experiments in genetics and cancer in the early 1900s. Today, the laboratory mouse is recognized as the preeminent mammalian model for modern genetic research. Mice are also used in a variety of other types of research, including cancer, immunology, toxicology, metabolism, developmental biology, diabetes, obesity, aging, and cardiovascular research. They are prized for many qualities, including their small size, short generation time, limited life span, and ease of breeding within the laboratory. The fact that they are genetically the best characterized of all mammals increases their value for all fields of study.

Mice belong to the order Rodentia, and most of the mice used in research belong to the genus *Mus*. Within that genus lies *Mus musculus*, with a number of subspecies described, including *Mus musculus domesticus* (the common house mouse) and *Mus musculus musculus*. Though laboratory mice originate from *Mus musculus* subspecies with a dash of other *Mus* species, they are typically referred to as *Mus musculus* or simply as "laboratory mice" (Didion and de Villena 2013).

DOI: 10.1201/9780429353086-1

stocks and strains

Both genetically diverse and genetically identical mice are used in research. Genetically diverse, or outbred, mice are called stocks, while genetically identical mice are called strains. Swiss Webster, ICR, and CD-1 are among the most commonly used genetically diverse stocks. There are thousands of genetically defined strains, including the following:

- **Inbred mice:** Mice of a particular inbred strain result from a minimum of 20 consecutive generations of brother × sister matings and are virtually identical to all other mice of the same strain; C57BL/6, BALB/c, C3H, FVB, 129, DBA, and CBA are among the most commonly used inbred strains.

- **Hybrid mice:** These mice are first-generation (F1) crosses between two different inbred strains. An example is the B6D2F1 line, which is a cross between the C56BL/6 (female parent) and DBA/2 (male parent) strains.

- **Recombinant inbred mice:** When F1 hybrids resulting from the same cross are mated together, the result is second-generation (F2) mice. Recombinant inbred strains result from 20 consecutive generations of brother × sister matings starting with the F2 mice.

- **Coisogenic strains:** These are strains of mice that differ from each other only at one gene, the difference being due to a spontaneous mutation that occurred in that gene. After the appearance of the mutation, the animals with the mutation are maintained as a separate strain from the original inbred strain. An example is DBA/2J-*Gpnmb*$^+$/SjJ, a mouse that does not develop the glaucoma seen in DBA/2 mice, but is otherwise genetically identical. (*Note:* Coisogenic strains can also be created using the technology for producing *knockout* mice; see below.)

- **Mice that carry induced mutations:** Mutations may be induced by chemicals (e.g., N-ethyl-N-nitrosourea), irradiation, or retroviruses. This category also includes the following categories:

  - **Transgenic mice:** These mice carry foreign DNA that was intentionally inserted into their own DNA. Examples include several transgenic mouse models of amyotrophic lateral sclerosis (Lou Gehrig's disease), all of which carry inserted copies of a human gene that codes for an abnormal enzyme (e.g., B6SJL-Tg(SOD1-G93A)1Gur).

- **Knockout and targeted mutation mice:** In these mice, a normal mouse gene has been rendered nonfunctional by the manipulation of embryos or stem cells by either *homologous recombination* or precision *endonucleases* such as CRISPR. Examples include knockout mice that are severely immunodeficient because they lack a gene that is necessary for the development of B and T lymphocytes (e.g., B6.129S7-*Rag1*^{tm1Mom}).

- **Congenic mice:** Congenic mice are similar to coisogenic mice, though the genetic dissimilarity between otherwise identical strains is created through breeding rather than as a result of a mutation. This is accomplished by breeding animals with the desired gene(s) to animals of a selected inbred host strain, then breeding their offspring back to the same host strain (backcrossing), and so on for at least ten generations. This process can be accelerated with selected genetic testing at each generation of backcrossing. Further, congenic mice have a much larger portion of the genome which differs from that of the host strain since surrounding genes are transferred along with the gene of interest during backcrossing.

nomenclature

Nomenclature is regulated by the International Committee on Standardized Genetic Nomenclature for Mice. Use of proper nomenclature for identifying mice is important, as the name of the mouse contains information that is essential to understanding its genetic makeup and differentiating it from the thousands of other stocks and strains of mice used in research. For the same reason, mouse nomenclature is complex, and understanding and using it may be challenging, although becoming familiar with the basics is helpful. The following guidelines represent a simplified overview of this topic. For a complete description of the rules and guidelines for nomenclature of both mice and their genes, refer to the Mouse Nomenclature

Home Page (www.informatics.jax.org/mgihome/nomen/index.shtml), located at the Mouse Genome Informatics web page (www.informatics .jax.org/).

Outbred mice are referred to by a laboratory (supplier) code followed by a stock designation. The stock designation consists of two to four uppercase letters. These letters may be followed by information in parentheses, indicating the origin of the stock. For example, Crl:CD1(ICR) is the outbred CD-1 mouse of ICR origin from Charles River Laboratories.

Inbred mice are referred to by a combination of letters (generally capitalized) and numbers, e.g., FVB, C3H, or 129. *Substrains* are colonies of the same inbred strain that have been separated for 20 generations. Generations pass on both sides, so researchers breeding commercially available mice while the vendor continues to breed the same mice is a common cause of substrain formation. Substrains are identified by a slash followed by a number and/or letters designating the institution/laboratory maintaining the colony. For example, C57BL/6N and C57BL/6J are two different substrains of the (now extinct) parental strain of C57BL. *Hybrid* mice are designated using the standard strain abbreviations (e.g., D2 for DBA/1 and B6 for C57BL/6) for the two parental strains. The female parent is always indicated first. B6D2F1 is a hybrid resulting from a mating between a female C57BL/6 mouse and a male DBA/2 mouse. *F1* indicates that the mouse is the first-generation offspring resulting from this mating. *Recombinant inbred* mice are designated by one- or two-letter strain abbreviations (e.g., D for DBA/2 and B for C57BL/6) for the progenitor strains separated by an "X" and with a strain number appended. For example, BXD25 is a recombinant inbred derived from C57BL/6 and DBA/2.

Transgenic mice are denoted using the standard nomenclature for the inbred strain (or hybrid) followed by the three-part designation for the inserted gene, e.g., B6SJL-Tg (SOD1-G93A)1Gur.

1. The first part of this designation (Tg) refers to the *mode of insertion* of the foreign gene (in this case, insertion by microinjection of the foreign DNA into one of the pronuclei of a fertilized egg).

2. The second part of the designation (SOD1-G93A) refers to the *foreign gene* that was inserted (in this case, a mutant form of the human *SOD1* gene).

3. The third part of the designation (1Gur) refers to the *founder line number* (in this case, "1") and the *registration code* assigned to the laboratory that produced the transgenic line (in this case, "Gur" for Dr. Mark Gurney).

Knockout or *targeted mutation* mice are designated in the same manner as mice with spontaneous mutations, along with an indication of the method by which the mouse gene of interest was rendered nonfunctional and a designation of the laboratory that produced the knockout. B6.129S7-*Rag1*^{tm1Mom} is a C57BL/6 mouse with a disruption of the *Rag1* gene. This disruption was the first founder line ("1") produced by a targeted mutation ("*tm*") in the laboratory of Drs. P. Mombaerts and S. Tonegawa ("*Mom*"). If that mouse had been produced using an endonuclease (e.g. CRISPR), the "*tm*" in the nomenclature would be replaced with "*em*".

Congenic mice are identified by the strain name of the background or host strain (often abbreviated) followed by a period, the name of the strain (often abbreviated) that donated the desired trait, a hyphen, and the gene symbol of the transferred gene. It is good to remember that any gene may be transferred in this manner—a "normal" mouse gene or a genetic modification such as a transgene or knocked-out gene. For example, B6.V-*Lep*^{ob} is a C57BL/6J mouse carrying the *obese* (*ob*) mutation in the *leptin* gene (*Lep*), which arose in the V strain.

anatomic and physiologic features

Important and unique anatomic and physiologic features of the mouse include the following:

- **Dentition**
  - The mouse is a monophydont hypsodont, meaning that it has one set of teeth (no baby teeth) that are open-rooted and continuously growing. The enamel contains iron, and the teeth primarily wear against each other.
  - This continuous growth can cause problems if the teeth are misaligned. This is called malocclusion and is seen more frequently in incisors.
  - Mice have eight teeth, and their dental formula is 1/1 incisors, 0/0 canines, 0/0 premolars, and 3/3 molars.

- **Skeleton**
  - The normal vertebral formula is C7,T13,L6,S4,C28.
  - The mouse normally has 13 pairs of ribs. The seven cranial pairs are *true* ribs and articulate with the sternum. In addition, there are six pairs of *false* ribs, the three most cranial of which connect to the caudal-most *true* rib, and the last three of which are *free* or *floating* with no attachment to other osseous structures.
- **External features**
  - Front and rear feet both have five digits each.
  - The female mouse normally has five pairs of nipples over the ventral thorax (three pairs) and abdomen (two pairs).
  - The male mouse has no nipples.
- **Hearing and vocalization**
  - Mice have a substantially narrower hearing range than humans (Heffner and Heffner 2007).
  - Typical of rodents, mice hear sounds of high frequency better than humans do.
  - Many commonly used inbred strains have age-related high-frequency hearing loss.
  - Ultrasonic vocalizations are believed to be used to facilitate or inhibit social interactions, alert to predator presence, express alarm or anxiety, and change if animals are in pain (Kurejova et al. 2010; Portfors and Perkel 2014).
- **Visual system**
  - Mouse retinas contain rods (light/dark sensing) and two types of cones (color sensing). Their cones have peak sensitivity at 360 nm (ultraviolet) and 508 nm (green) (Wang et al. 2011). They cannot see the color red, which is why so much mouse cage furniture is red; humans can see through it, but it is opaque to mice.
  - Many strains of mice are blind from weaning on, due to the fixation of a retinal degeneration allele in the initial population (Wong and Brown 2006). Many more albino strains lose sight rapidly due to vision damage from lights (De Vera Mudry et al. 2013).

- Mice are nearsighted when compared with humans, but are three to five times more sensitive to motion in their visual field (Huberman and Niell 2011).

- **Gastrointestinal system**
  - The alimentary canal is similar to other mammals (except ruminants) and consists of the esophagus, stomach, duodenum, jejunum, ileum, cecum, colon, and rectum.
  - The large cecum is where the mouse digests a great deal of the plant matter it ingests.
  - The stomach is divided into cardiac (nonglandular) and pyloric (glandular) sections. The nonglandular portion is lined by squamous epithelium.
  - The mouse cannot vomit.
  - The mouse does not have an appendix.

- **Urogenital system**
  - The right kidney is normally anterior to the left kidney.
  - In males, the inguinal canal remains open, and the testes may be retracted into the abdominal cavity. Males typically have an *os* penis, a small bone over the urethra near the tip of the penis. *Preputial glands* are paired structures that lie subcutaneously near the tip of the prepuce. Occasionally, these glands become inflamed and will present as a small mass alongside the prepuce.
  - Males have several accessory reproductive glands located within the abdomen, including the prostate, the seminal vesicles, and the coagulating glands. Secretions from these glands comprise the copulation plug, deposited in the vagina by the male after mating.
  - In females, the reproductive tract includes two uterine horns that combine to form the *median corpus*. The *clitoral glands* lie subcutaneously just lateral to the opening of the urethra. As with the preputial glands, the clitoral glands will occasionally become inflamed.
  - Males are distinguished from females by the scrotal sac containing the testes and by a longer anogenital distance (Figure 1.1).
  - The placenta of the mouse is hemochorial.

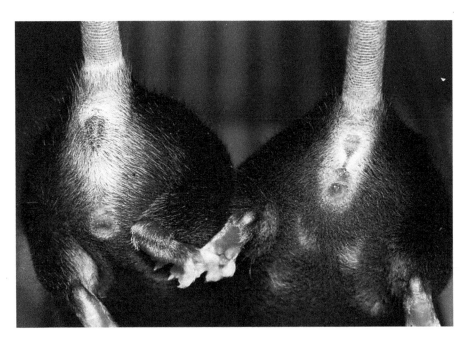

Figure 1.1 Difference in anogenital distance between male (left) and female (right) mice. The distance between the anus and the external genitalia is shorter in females than in males.

- The urine is normally clear, yellow, and quite concentrated (up to 4.3 osmol/kg) (Silverstein 1961). Urine specific gravity typically is in the region of 1.040 and can approach 1.080. Mouse urine specific gravity exceeds the limits of some dipstick and refractometer measurements, and so veterinary refractometers with range up to 1.080 are more useful. Large amounts of protein are normally excreted in the urine of mice, including uromucoid, alpha and beta globulins, and major urinary protein, the primary mouse pheromone carrier and allergen. The pH of normal mouse urine is approximately 5.0.
- Maternal immunoglobulin is transferred to the pups across the placenta and across the intestinal epithelium from colostrum and milk for 16 days after parturition.
- **Respiratory system**
  - The mouse has one lung lobe on the left side and four lobes (cranial, middle, caudal, and accessory) on the right.

- **Glands associated with the eye**
  - The *Harderian* gland is horseshoe-shaped and located within the orbit. It produces a secretion that lubricates the eyelids.
  - The *extraorbital* gland is located subcutaneously just ventral and anterior to the ear. It produces a secretion that lubricates the globe.
  - The *intraorbital* gland is found near the lateral canthus of the eye and produces a secretion that lubricates the globe.
- **Spleen**
  - Black spots in spleens of pigmented mice, e.g., C57L/6, often are melanin pigment, an incidental finding

normative values

Basic Biologic Parameters

Typical values for miscellaneous biologic parameters are presented in Table 1.1. (*Note*: Significant variation of values may occur between individual mice, strains and stocks, laboratories, and methods of sampling. It is imperative that individual laboratories establish normal values for their specific facility.)

Clinical Chemistry

For clinical chemistry, some values are measured in serum, which is the fluid from clotted blood, and other values are measured from

TABLE 1.1: MISCELLANEOUS BIOLOGIC PARAMETERS OF THE MOUSE

Parameter	Typical Value
Diploid chromosome number	40
Life span	2–3 years
Adult body weight	20–40 g
Body temperature	36.5–38.0°C (97.5–100.4°F)
Metabolic rate	180–505 kcal/kg/day
Food intake	12–18 g/100 g body weight/day
Water intake	15 ml/100 g body weight/day
Respiratory rate	80–230 breaths/min
Heart rate	500–600 beats/min

plasma, fluid from unclotted blood. Various anticoagulants are used to keep the blood from clotting, and the anticoagulant used should be tailored to the assay. Approximate values for clinical chemistry parameters are shown in Table 1.2. The values represent ranges in mean values reported for mice between 1 and 12 months of age. The reference values provided in Table 1.2 are based on values from various strains, ages, sex, and laboratory conditions. They may provide

TABLE 1.2: CLINICAL CHEMISTRY VALUES OF THE MOUSE*

Analyte Evaluated	Reported Values (US units)	Reported Values (SI units)	Tissue/System/Process
Glucose	97–239 mg/dl	5.38–13.27 mmol/l	Pancreas (diabetes), liver
Blood urea nitrogen (BUN)	18–35 mg/dl	6.43–12.5 mmol/l	Kidney
Creatinine	0.2–0.7 mg/dl	15.3–53.4 mmol/l	Kidney
Sodium	135–165 mEq/l	135–165 mmol/l	Electrolyte/water balance
Potassium	4.9–7.1 mEq/l	4.9–7.1 mmol/l	Electrolyte/water balance
Chloride	99–128 mEq/l	99–128 mmol/l	Electrolyte/water balance
Calcium	8.8–12.2 mg/dl	2.2–3.05 mmol/l	Thyroid/parathyroid, intestine, pancreas, kidney, bone metastasis
Phosphorus	5.7–14.9 mg/dl	1.8–4.8 mmol/l	Kidney
Iron	110–252 mcg/dl	16.7–45.1 umol/l	Iron transport and storage
Alanine transaminase (ALT or SGPT)	0–228.2 IU/l	0–228.2 U/l	Liver
Aspartate transaminase (AST or SGOT)	35–137.7 IU/l	35–137.7 U/l	Liver, heart, skeletal muscle
Alkaline phosphatase (ALP)	44–118 IU/l	44–118 U/l	Liver, GI tract, kidney, bone
Lactic dehydrogenase (LDH)	400–800 IU/l	400–800 U/l	Liver, heart, skeletal muscle, LDH-elevating virus infection
Total protein	4.9–7.3 g/dl	49–73 g/l	Liver function, immunoglobulin status
Albumin	2.0–4.7 g/dl	20–47 g/l	Liver function
Globulin	1.2–2.8 g/dl	12–28 g/l	Liver function
Cholesterol	57.7–123.8 mg/dl	1.5–3.2 mmol/l	Cardiovascular disease
Triglycerides	75.6–198.5 mg/dl	0.85–2.23 mmol/l	Cardiovascular disease
Total bilirubin	0–0.5 mg/dl	0–8.55 mmol/l	Heme catabolism, cholestasis

* Values from Kurtz and Travlos (2018), Serfilippi et al. (2003), Otto (2016), Luong (2018). Values are for C57BL/6 mice of various ages; the values given are +/- 2 SD from the mean.

a useful starting point, but individual laboratories should attempt to establish specific normal values based on these variables.

Urinalysis

Evaluation of mouse urine is complicated by the small volumes that are usually available (Kurien et al. 2004). For studies that require multiple or quantitative urinalyses, 24-h urine collections are usually obtained by using metabolic cages as described in Chapter 6. Increased drinking (polydipsia) and increased urination (polyuria) are typical of diabetes and some types or stages of renal disease. Sick mice, like other sick animals, may drink less than normal, resulting in decreased urine output. In addition, treatment of drinking water with chlorine, acid, or research-related compounds can affect palatability of water and may reduce water intake and urine output. Some mice chew and grind food, resulting in the disappearance of more food than is ingested. Much of this ground food, along with feces, can contaminate urine samples, resulting in abnormal urine sediment, and bacterial growth that can alter protein and glucose values. Finally, small volumes of liquid are susceptible to evaporation because of their relatively high surface area, and so most 24-h collections of mouse urine are prone to evaporation artifact.

The common laboratory evaluations involved in urinalysis are for color, specific gravity, protein, glucose, and evaluation of sediment. Typical values are shown in Table 1.3 (note that some values in this table are provided per 24 h).

Urine sediment is the nonliquid material that remains after urine is centrifuged. This material is evaluated microscopically for the presence of cells, casts, crystals, and bacteria. With 24-h urine specimens, it is not uncommon to find contaminating plant material (from

TABLE 1.3: NORMAL PARAMETERS OF URINE IN THE MOUSE*

Parameter	Approximate Normal Value
Output per day	1/3 2.0 mL
Color	Clear or slightly yellow
Volume	0.5–2.5 mL/24 h
Specific gravity	1.040
pH	5.0
Glucose	0.5–3.0 mg/24 h
Protein	0.6–2.6 mg/24 h

* Values are from Kurtz (2018), Watts (1971), Luong (2018).

ground feed), bacteria, and pinworm eggs or gastrointestinal proto-zoa (in infested colonies, from fecal contamination). *Klossiella muris* is a protozoal parasite that can infect mouse kidneys and might be discerned in urine specimens, but it is very rare in modern, well-maintained mouse colonies.

hematology

Hematology is the study of blood and usually refers to the study of its cellular components, including erythrocytes or red blood cells (RBCs), leukocytes or white blood cells (WBCs), and platelets. Blood can be analyzed with automated equipment (automated complete blood count) and by microscopic examination of stained blood smears.

Hematology values can vary with mouse strain/stock, source, age, sex, blood sampling method, environmental conditions, pathogen status, and laboratory. The reference values provided in Table 1.4 are based on values from various strains, ages, sex, and laboratory conditions. They may provide a useful starting point, but individual laboratories should attempt to establish specific normal values based on these variables.

TABLE 1.4: TYPICAL HEMATOLOGIC VALUES OF THE MOUSE*

Parameter	Reported Mean Values	Units
Hematocrit	39.2–54.4	%
Red blood cell number	9.17–12.1	10^6/uL
Hemoglobin concentration	13.1–17.9	g/dl
MCV	38.4–	fL
MCH	12.9–16.8	pg
MCHC	27–37.3	g/dl
Platelets	445–1815	10^3 platelets/μL
White blood cells	3.63–16.7	10^3 cells/μL
Neutrophils	0.9–5.24	10^3 cells/μL
Eosinophils	0.06–0.69	10^3 cells/μL
Basophils	0.006–0.093	10^3 cells/μL
Lymphocytes	2.03–13.3	10^3 cells/μL
Monocytes	0.47–0.61	10^3 cells/μL

* Values are a range of the means of data collected from 6-month old male and female mice of 30 different inbred strains, housed at the Jackson Laboratory (Peters et al. 2007). Reported ranges and standard deviations for some parameters are very large.

immunodeficient mice

Some studies, such as those involving certain infectious pathogens or transplantation of tissue from another species, require that the host animal is immunodeficient; that is, that the animal has an immune system that does not function sufficiently to eliminate foreign microbes or tissues. Though it is possible to create such animals by treatments such as irradiation to deplete bone marrow precursors of immune effector cells, a number of mouse genetic mutants exist that have various immunologic impairments that render them immunodeficient (Belizario 2009; Shultz et al. 2019). Some of these mutations have occurred spontaneously, while others have been experimentally produced. These mutants are often produced on common background strains such as BALB/c and C57BL/6. Efforts to cross-breed individuals of different immunodeficiency types have produced offspring with combinations of these immunologic defects; thus, a library of mice with defects representing a broad range of immunodeficiency disorders is available. The more commonly encountered mutants are described here.

Nude mice lack a developed thymus and are, therefore, sometimes referred to as *athymic*. A single-gene mutation at *Foxn1* is responsible for both the lack of a thymus and a general lack of body hair in homozygotes (Figure 1.2), the latter of which has resulted in the

Figure 1.2 Adult nude mouse. The hairlessness is a feature coincident with the lack of a thymus in the nude mouse.

nickname *nude*. Because the thymus is needed for maturation of T lymphocytes, athymic mice have impaired immune responses that depend on T lymphocytes, including defects in antibody formation, graft rejection, killing of virus-infected or tumor cells, and delayed-type hypersensitivity. In addition, nude mice are characterized by an antibody response that is confined to the IgM class, and an increased natural killer cell response.

Severe combined immunodeficiency (SCID) mice possess a genetic autosomal recessive mutation, *Prkdc*^{scid}, a defect that impairs the production of both T and B lymphocytes. In contrast to nude mice, SCID mice are haired. With age, some SCID strains demonstrate low levels of circulating antibody, a phenomenon known as *SCID leakiness*.

Nonobese diabetic (NOD) mice serve as a model for type 1 diabetes, but also serve as a good model for autoimmune disease and the understanding of immune tolerance. NOD mice have a unique major histocompatibility complex (MHC) haplotype (*H2*^{g7}), which is an essential contributor to disease susceptibility. Female NOD mice develop diabetes at a higher rate and earlier onset than males.

Recombination activating gene (Rag) mice have a deletion of either *Rag1* or *Rag2*. Because this defect results in the arrest of T and B lymphocyte receptor rearrangement, they lack T and B cell differentiation, and therefore T and B lymphocytes. Rag mice have high levels of natural killer cell activity, but do not demonstrate antibody leakiness as seen in SCID mice.

Beige mice have an autosomal recessive mutation, *Lyst*^{bg}, which results in defective natural killer cells. The reference to beige indicates that this mutation also affects coat color in pigmented mice, which is how it was discovered. Beige mice have a number of immunological perturbations, including reduced bactericidal activity of granulocytes and severe deficiency of natural killer cells.

NOD.SCID mice have both the NOD and SCID mutations, though they do not develop diabetes. Human cells can be more easily engrafted into NOD.SCID mice than into SCID mice, as they have reduced natural killer cell activity compared with SCID mice. These mice can be implanted with human fetal thymic and liver tissue, with resulting extensive infiltration of organs and tissues with human cells.

NOD-SCID-gamma (NSG®, NOG, NCG) mice are NOD mice carrying both the *scid* allele and a knockout of the Il2 receptor gamma gene. This means they lack functional or mature T, B, and NK cells as well as having reduced macrophage and dendritic cell function, which allows for reconstitution of human immune systems or implantation of human tumors. The *Il2rg* knockout has also been crossed onto a NOD-*Rag* mouse (NRG) resulting in a similar genotype. The mice listed above are not necessarily interchangeable, as their NOD genetic background differs, and the alleles they carry have been placed on the NOD background by classic backcrossing or by targeted mutation. These severely immunodeficient mice are the primary humanized mice used today.

wild mice

Some studies require the use of mice other than the domesticated *Mus musculus* subspecies. These include *Mus spretus* and various species of *Peromyscus* species (Figure 1.3). Often, the microbial status of such animals is uncertain, and appropriate strategies must be employed to contain infectious agents that may pose harm to other animals or personnel. For example, one study of wild mice on a university campus found evidence of infection with a variety of potential murine pathogens, including mouse hepatitis virus and *Helicobacter*, and infestation with pinworms and ectoparasites (Parker et al. 2009).

Mus spretus, also referred to as the western Mediterranean mouse, are smaller than the standard laboratory *Mus musculus* and have a white-bellied agouti coat. The species is distributed in northern Africa, throughout Portugal, and across all but the northern fringe of Spain into the southern region of France, and several inbred lines have been developed from wild-trapped ancestors. *M. spretus* that

Figure 1.3 An adult female deer mouse.

have been inbred and kept in captivity are more aggressive than laboratory mice and retain many other wild-derived traits. *Mus spretus* diverged from *M. musculus* over 1 million years ago, and the two species are genetically and phenotypically divergent. Due to the overall reduced genetic polymorphism, and consequent phenotypic polymorphism, seen in laboratory mice, *M. spretus* serves as a reservoir of additional phenotypic variation; thus they have become a valuable complement to laboratory mice for studying the genetics of complex traits (Dejager et al. 2009).

Unlike *Mus*, *Peromyscus* species are native to North and South America and include *P. maniculatus*, which is often referred to as the deer mouse. There is great variation in size, though species most commonly used in the laboratory are typically the approximate size of laboratory mice. There are approximately 55 distinct *Peromyscus* species, and some of these are used in research. Captive colonies are commercially available, and the Peromyscus Genetic Stock Center (www://sc.edu/study/colleges_schools/pharmacy/centers/perom yscus_genetic_stock_center/index.php) at the University of South Carolina is a source of various species of *Peromyscus* for investigators. Deer mice are the most common *Peromyscus* used in research and have been utilized in ecology and epidemiology studies (Bedford and Hoekstra 2015). Because the deer mouse is a carrier of hantavirus and has been implicated in transmission of the pathogen to humans, it has been used in relevant studies; however, it is that same feature that requires caution when maintaining wild-caught deer mice.

Although wild mice are generally very adaptable and thrive in conditions and caging similar to those used for care of domesticated laboratory mice, housing and care of wild mice should include provisions that meet the special features of the animals. For example, though evidence exists to suggest that wild mice may not be an important source of infectious agents for laboratory mice (Parker et al. 2009; Dyson et al. 2009), it is generally assumed that wild-caught mice may harbor infectious agents that pose a threat to other animals or personnel. For that reason, some facilities choose to take prudent precautions, such as use of added personal protective equipment (e.g., filtered masks, disposable gowns), use of filtered caging and separation from other animals, and autoclaving of contaminated caging and bedding. Further, it should be noted that these species are more athletic than laboratory mice and adept at escape. Deer mice, for example, are skilled jumpers and climbers, and care must be taken to prevent escape when the cage is opened. As with most mice, wild mice thrive when provided with abundant bedding and nesting material.

behavior, well-being, and enrichment

Behavior

In general, the domestic mouse is not terribly aggressive toward handlers and will attempt to evade rather than confront. The handler is most at risk when attempting to grasp or restrain the animal, as this may cause the animal to attempt biting. Occasionally, a mouse will bite a handler who is reaching toward the animal. Between strains, there may be marked behavioral differences, and this is an important factor in studies designed to evaluate behavioral phenotypes (Bothe et al. 2005). Abnormal behavior may signal underlying disease and should be reported to veterinary staff.

Although a social species, mice will show *aggressive behavior* with one another and may inflict serious injuries. This is particularly true with males and with some strains, such as the SJL or BALB/c. Fighting may be related to establishment of hierarchy and defense of territory (Theil et al. 2020), and scuffles are commonly seen after cage changes, although this can be mitigated by transferring nesting material at cage change. A more common observation in mice, however, especially mice on a C57BL/6 background, is *barbering*, a focal loss of hair or whiskers with no wounds and a very sharp margin between the areas of hair loss and normal hair (Figure 1.4). Barber mice can barber others or self-barber. The behavior is a compulsive pulling of hair and unrelated to dominance status in the cage (Garner et al. 2004). In spite of these behaviors, it should be understood that mice are social animals and are best maintained in compatible groups (Blankenberger et al. 2018).

Mice are most active during the dark phase and are therefore classified as *nocturnal*. In the laboratory environment, activity is also noted during the light phase as well. Mice sleep in fragmented bouts, not in long stretches, and so routine husbandry disturbances during the light phase are less disruptive than previously thought (Robinson-Junker et al. 2019). Mice commonly demonstrate burrowing and nesting behavior, and therefore abundant bedding and other material that encourages such behavior should be supplied.

Figure 1.4 Barbering on the ventrum of a C57BL/6 mouse. Note the sharp demarcation between the area with hair loss and the normal hair, which is characteristic of barbering.

As an animal with a high body mass to surface area ratio, these behaviors also serve to help the mouse maintain body temperature. Because they have limited means to regulate their temperature (cannot pant, very few sweat glands and those present only on the feet), mice must behaviorally avoid hyperthermia, primarily by sleeping during warmer parts of the day and being active at night. In addition, the relatively long, hairless tail serves as a conduit for heat exchange.

Mice spend a great deal of their time grooming themselves and others. Self-grooming behavior in mice is a fixed action pattern, meaning that it proceeds in a similar way from mouse to mouse, but a disruption of this pattern may be an indication that mice will proceed to ulcerative dermatitis, a non-infectious mouse condition with a behavioral component (Adams et al. 2016). Juvenile mice do seem to exhibit play behavior (Terranova and Laviola 2005), although it is not as easily recognized by humans as such, in comparison with rats.

Unlike many other domesticated animals, one of the chief goals of keeping laboratory mice is making more of them (Pritchett and Taft 2007). Breeding and maternal behaviors are commonly observed in the laboratory setting, and disruption of these behaviors can frustrate research goals. Mating occurs in the dark part of the light cycle, during late proestrus. The actual mating event takes place over 15–60 minutes, and the deposition of a copulation plug by the male indicates a successful mating event, although not always pregnancy. Male mice have a copulation clock that is switched on after a mating event, resulting in a decrease in infanticide if presented with pups 18–22 days after copulation (Perrigo et al. 1992). Pregnant female mice do not significantly alter their behavior until close to parturition, at which point nest building behavior increases, and aggression toward strange males dramatically increases. Gestation length in mice is a function of both litter size and genetic background (Murray et al. 2010). Parturition typically occurs during the dark phase and lasts for approximately 1 h, with birth occurring 2–3 h after the first appearance of fluid at the vaginal opening. Despite received mouse wisdom, infanticide is not common in stably housed breeding groups, and "cannibalism" is a means of disposal of pups either born dead or dying shortly after birth (Brajon et al. 2021). Maternal behavior in mice includes nursing, licking/grooming of pups, and retrieving pups to the nest (Brown et al. 1999) and varies by strain or stock (some genetic backgrounds exhibit more of one type of maternal behavior than others) (Champagne et al. 2007).

Well-being

Mice can show two general types of problematic behavior. One type is abnormal behavior such as stereotypies. The second type is normal behavior for the mouse that is problematic in captivity, such as aggression. The second type, although a welfare concern, is inherent to mice, and management should be considered, since the behavior cannot be eradicated. Stereotypies are repetitive and invariant patterned behaviors with no apparent goal or function. Current thinking attributes most stereotypies to the expression of stress related to frustrated natural behaviors or drives (Garner 2005), but the presence of a stereotypical behavior does not necessarily indicate a negative state of well-being (Novak et al. 2016). Mouse stereotypies tend to be motion-related, since domesticated mice retain natural drives to explore and patrol a territory. They may also be related to digging, another retained wild mouse behavior, or to attempts to escape the cage. Some common stereotypies described in mice include: twirling, circling, route-tracing, bar-mouthing, tail-carrying, and corner jumping (Garner 2022). Some investigators have suggested that the use of running wheels by mice may constitute a stereotypy, but most feel that the use of running wheels normalizes mouse biology and behavior (Walker and Mason 2018).

Treatment of stereotypies is difficult, if not impossible, and the best treatment is prevention, something that is difficult in standard laboratory housing (Bailoo et al. 2020), although group housing does decrease levels of stereotypic behavior. Providing animals in standard laboratory housing ways to express the full repertoire of natural behaviors remains a challenge to be tackled by researchers, veterinarians, and producers of caging alike.

Enrichment

Happily for both mice and science, the importance of enrichment to the welfare, development, and use of mice as a model system has been widely recognized (Bayne 2018; Bayne and Würbel 2014). Building a nest is crucial to the welfare and development of mice, and, as such, nesting material should not be considered enrichment, but rather a basic need on the same level as food and water (Figure 1.5). A larger cage with more territory to explore can be enriching to mice. Some mouse housing companies are building these opportunities into their caging systems with provision of

Figure 1.5 Mouse at rest within nest built from added paper nesting material.

tubes that allow the connection of cages. Further enrichment of a standard cage might include shelters such as huts or tubes—useful for animals to nest in or climb on to self-rescue from floods or to transfer animals from cage to cage. It might also include added food, such as sunflower seeds or mealworms, which can be sourced irradiated or certified and either added to the food hopper or distributed in the substrate to encourage foraging. Mice can also benefit from additional gnawing substrates such as wood blocks or nylon objects, which may show some benefit in reducing food grinding behavior. Inveterate climbers, mice will use also climbing areas such as the wire lids of their cages or other climbing opportunities. Finally, running wheels provide both enrichment and exercise for mice. There are products that can be placed within typical cages and also provide a shelter for animals. Before changing enrichment strategies, consult with facility management, veterinarians, and investigators to be sure that the chosen enrichment will not interfere with biosecurity or study goals (Perrigo et al. 1992).

references

Adams SC, Garner JP, Felt SA, Geronimo JT, Chu DK. 2016. A "pedi" cures all: toenail trimming and the treatment of ulcerative dermatitis in mice. *PLoS One* **11**:e0144871.

Bailoo JD, Voelkl B, Varholick J, Novak J, Murphy E, Rosso M, Palme R, Wurbel H. 2020. Effects of weaning age and housing conditions on phenotypic differences in mice. *Sci Rep* **10**:11684.

Bayne K. 2018. Environmental enrichment and mouse models: current perspectives. *Animal Model Exp Med* **1**:82–90.

Bayne K, Würbel H. 2014. The impact of environmental enrichment on the outcome variability and scientific validity of laboratory animal studies. *Rev Sci Tech* **33**:273–280.

Bedford NL, Hoekstra HE. 2015. *Peromyscus* mice as a model for studying natural variation. *eLife* **4**:e06813.

Belizario JE. 2009. Immunodeficient mouse models: an overview. *Open Immunol J* **2**:79–85.

Blankenberger WB, Weber EM, Chu DK, Geronimo JT, Theil J, Gaskill BN, Pritchett-Corning K, Albertelli MA, Garner JP, Ahloy-Dallaire

J. 2018. Breaking up is hard to do: does splitting cages of mice reduce aggression? *Appl Anim Behav Sci* **206**:94–101.

Bothe GW, Bolivar VJ, Vedder MJ, Geistfeld JG. 2005. Behavioral differences among fourteen inbred mouse strains commonly used as disease models. *Comp Med* **55**:326–334.

Brajon S, Morello GM, Capas-Peneda S, Hultgren J, Gilbert C, Olsson A. 2021. All the pups we cannot see: cannibalism masks perinatal death in laboratory mouse breeding but infanticide is rare. *Animals* **11**:2327.

Brown RE, Mathieson WB, Stapleton J, Neumann PE. 1999. Maternal behavior in female C57BL/6J and DBA/2J inbred mice. *Physiol Behav* **67**:599–605.

Champagne FA, Curley JP, Keverne EB, Bateson PP. 2007. Natural variations in postpartum maternal care in inbred and outbred mice. *Physiol Behav* **91**:325–334.

De Vera Mudry MC, Kronenberg S, Komatsu S, Aguirre GD. 2013. Blinded by the light: retinal phototoxicity in the context of safety studies. *Toxicol Pathol* **41**:813–825.

Dejager L, Libert C, Montagutelli X. 2009. Thirty years of *Mus spretus*: a promising future. *Trends Genet* **25**:234–241.

Didion JP, de Villena FP. 2013. Deconstructing *Mus gemischus*: advances in understanding ancestry, structure, and variation in the genome of the laboratory mouse. *Mamm Genome* **24**:1–20.

Dyson MC, Eaton KA, Chang C. 2009. *Helicobacter* spp. in wild mice (*Peromyscus leucopus*) found in laboratory animal facilities. *J Am Assoc Lab Anim Sci* **48**:754–756.

Garner JP. 2005. Stereotypies and other abnormal repetitive behaviors: potential impact on validity, reliability, and replicability of scientific outcomes. *ILAR J* **46**:106–117.

Garner JP, Dufour B, Gregg LE, Weisker SM, Mench JA. 2004. Social and husbandry factors affecting the prevalence and severity of barbering ('whisker trimming') by laboratory mice. *Appl Anim Behav Sci* **89**:263–282.

Garner JP, Gaskill B, Rodda C, Dufour B, Prater A, Klein J, Wurbel H, Mason G, Olsson A, Weber EM, Geronimo JT, May C. 2022. Mouse ethogram: an ethogram for the laboratory mouse. Available at: https://mousebehavior.org/stereotypy/.

Huberman AD, Niell CM. 2011. What can mice tell us about how vision works? *Trends Neurosci* **34**:464–473.

Keeler CE. 1931. *The Laboratory Mouse: Its Origin, Heredity, and Culture.* Cambridge, MA: Harvard University Press.

Kurejova M, Nattenmüller U, Hildebrandt U, Selvaraj D, Stösser S, Kuner R. 2010. An improved behavioural assay demonstrates that ultrasound vocalizations constitute a reliable indicator of chronic cancer pain and neuropathic pain. *Mol Pain* **6**:18.

Kurien BT, Everds NE, Scofield RH. 2004. Experimental animal urine collection: a review. *Lab Anim* **38**:333–361.

Kurtz DM, Travlos GS, Eds. 2018. *The Clinical Chemistry of Laboratory Animals.* Boca Raton: CRC Press.

Luong RH. 2018. The laboratory mouse, pp. 1–31. In *The Clinical Chemistry of Laboratory Animals*, Kurtz DM, Travlos GS, Eds. Boca Raton: CRC Press.

Murray SA, Morgan JL, Kane C, Sharma Y, Heffner CS, Lake J, Donahue LR. 2010. Mouse gestation length is genetically determined. *PLoS One* **5**:e12418.

Novak J, Bailoo JD, Melotti L, Wurbel H. 2016. Effect of cage-induced stereotypies on measures of affective state and recurrent perseveration in CD-1 and C57BL/6 mice. *PLoS One* **11**:e0153203.

Otto GP, Rathkolb B, Oestereicher MA, Lengger CJ, Moerth C, Micklich K, Fuchs H, Gailus-Durner V, Wolf E, de Angelis MH. 2016. Clinical chemistry reference intervals for C57BL/6J, C57BL/6N, and C3HeB/FeJ mice (*Mus musculus*). *J Am Assoc Lab Anim Sci* **55**:375–386.

Parker SE, Malone S, Bunte RM, Smith AL. 2009. Infectious diseases in wild mice (*Mus musculus*) collected on and around the University of Pennsylvania (Philadelphia) Campus. *Comp Med* **59**:424–430.

Perrigo G, Belvin L, Vom Saal FS. 1992. Time and sex in the male mouse: temporal regulation of infanticide and parental behavior. *Chronobiol Int* **9**:421–433.

Peters LL, Schultz D, Godfrey D. 2007. Aging study: blood hematology in 30 inbred strains of mice. MPD:Peters 4. Mouse Phenome Database web resource (RRID:SCR_003212). Bar Harbor: The Jackson Laboratory. https://phenome.jax.org 18 September 2022.

Portfors CV, Perkel DJ. 2014. The role of ultrasonic vocalizations in mouse communication. *Curr Opin Neurobiol* **28**:115–120.

Pritchett KR, Taft RA. 2007. Reproductive biology of the laboratory mouse, pp. 91–122. In *The Mouse in Biomedical Research: Normative Biology, Husbandry, and Models*, vol **3**, Fox J, Barthold S, Davisson M, Newcomer C, Quimby F, Smith A, Eds. New York: Academic Press.

Robinson-Junker A, O'Hara B, Durkes A, Gaskill B. 2019. Sleeping through anything: the effects of unpredictable disruptions on mouse sleep, healing, and affect. *PLoS One* **14**:e0210620.

Serfilippi LM, Stackhouse Pallman DR, Russell B, Spainhour CB. 2003. Serum clinical chemistry and hematology reference values in outbred stocks of albino mice from three commonly used vendors and two inbred strains of albino mice. *J Am Assoc Lab Anim Sci* **42**:46–52.

Shultz LD, Keck J, Burzenski L, Jangalwe S, Vaidya S, Greiner DL, Brehm MA. 2019. Humanized mouse models of immunological diseases and precision medicine. *Mamm Genome* **30**:123–142.

Silverstein E. 1961. Urine specific gravity and osmolality in inbred strains of mice. *J Appl Physiol* **16**:194.

Terranova ML, Laviola G. 2005. Scoring of social interactions and play in mice during adolescence. *Curr Protoc Toxicol* Chapter 13:Unit13.10.

Theil JH, Ahloy-Dallaire J, Weber EM, Gaskill BN, Pritchett-Corning KR, Felt SA, Garner JP. 2020. The epidemiology of fighting in group-housed laboratory mice. *Sci Rep* **10**:16649.

Walker M, Mason G. 2018. A comparison of two types of running wheel in terms of mouse preference, health, and welfare. *Physiol Behav* **191**:82–90.

Wang YV, Weick M, Demb JB. 2011. Spectral and temporal sensitivity of cone-mediated responses in mouse retinal ganglion cells. *J Neurosci* **31**:7670–7681.

Watts RH. 1971. A simple capillary tube method for the determination of the specific gravity of 25 and 50 micro 1 quantities of urine. *J Clin Pathol* **24**:667–668.

Wong AA, Brown RE. 2006. Visual detection, pattern discrimination and visual acuity in 14 strains of mice. *Genes Brain Behav* **5**:389–403.

husbandry

Good husbandry is imperative for the optimal performance and health of mice in research, education, or testing. It involves attention to all aspects of the animals' environment, including both the macroenvironment (the room, cubicle, etc., in which the cage is kept) and the microenvironment (within the cage itself). It also involves attention to sanitation, nutrition, and transportation of animals into and out of the facility. For many facilities, knowledge of breeding systems is essential, as is a program of genetic quality control. Accurate identification of mice and carefully maintained records are vital for all facilities.

housing

Caging

Cages for laboratory mice should be made of a nonporous, easily sanitizable material that is comparatively resistant to impact and can withstand frequent exposure to hot water and detergent in the cage washer. Nonopaque caging is preferred in most cases as it allows reasonable visualization of mice, feed, bedding, and water without opening the cage. If caging will be autoclaved—recommended for high-health status mice and high-risk mice such as severely immunodeficient animals—the material must be able to withstand autoclaving. Most mouse cages used to house mice are constructed of plastic, as plastics can meet all of the above criteria; however, some cage components, such as stainless steel wire bar lids, are made

DOI: 10.1201/9780429353086-2

of other materials. Table 2.1 lists the plastics most commonly used for mouse cages, along with some of their advantages and disadvantages. For ease of cleaning, it is preferable to have cages with rounded corners that minimize the packing of soiled bedding.

Cage bottoms may be either solid or wire grid. Wire-bottom cages are not provided with bedding, and they allow urine and feces to fall through onto a cage pan located below the cage. While they are advantageous for some species and may be necessary for some studies with mice (e.g., toxicity testing), they are generally not recommended for mice as they can interfere with thermoregulation and breeding performance (National Research Council 2011).

There are many different types of cage tops. Typically, mouse cages are covered with a wire lid that contains a holder for feed and that may or may not have a place to put a water bottle. It is important that the bars or strands of wire be placed close enough together, or woven in a tight enough mesh, that mice cannot escape through, or become trapped in the openings. This is a particular problem with pre-weanling/weanling mice and with wild-derived mice of all ages. It is astonishing how small an opening a determined mouse can squeeze through; as a general rule of thumb, an average-sized mouse can slip through any opening large enough to insert a pencil. Wire lids are not recommended for animals with cranial implants, as these may be damaged by the wire. The wire lid may then be covered with some type of a filter top to minimize the spread of microorganisms between

TABLE 2.1: PLASTICS COMMONLY USED FOR MOUSE CAGES

Material	Clarity	Impact/Heat Resistance	Chemical Resistance	Chemical Resistance
Polypropylene	Variable	High	High	Medium
Polystyrene	Transparent	Low	Low	Low
Polycarbonate[a]	Transparent	Very high	Generally high[b]	Medium
High-temperature polycarbonate	Transparent, amber tint, red tint, blue tint	High	Generally high[b]	High
Polyethylene	Opaque	Medium	Very high	Low
Polyetherimide	Transparent, dark amber tint	Low	High	Very high
Polysulfone[c]	Transparent, amber tint	High	High	High

[a] May leach bisphenol A (BPA) as caging ages (Howdeshell et al. 2003).
[b] Susceptible to damage by strong alkaline agents.
[c] *Damaged polysulfone cages may leach bisphenol (Gorence et al. 2019).*

cages. There are many different types of filter tops, a typical example being the microisolator top shown in Figure 2.1. Filter tops should be sufficiently separated from the wire cover so that the mice cannot chew them or fitted with metal protection for animals that cannot be housed with wire lids. All filter tops will interfere to some extent with the ventilation of the cage, thereby influencing the microenvironment within the cage (Lipman 1999). For example, the use of filter tops will generally lead to increases in the intracage temperature, ammonia concentration, and humidity.

Disposable cages—cages that are discarded after a single use—have been on the market for many years but, until recently, were of limited usefulness because of their extreme fragility. More recently, disposable caging made of polyethylene terephthalate—abbreviated as PET or PETE—has become commercially available. PET cages are lightweight, clear, impact resistant, and recyclable. While PET cannot be autoclaved, it can be sterilized with ethylene oxide gas or gamma irradiation. PET caging is available for static housing conditions or for use in ventilated caging systems. Disposable cages made of recyclable polypropylene are also available. Disposable caging is an increasingly popular choice for containment situations, e.g., quarantine, or studies using infectious agents or toxic chemicals. It is also useful in situations where cage wash equipment is not available. Extending the use of disposable cages may be possible with

Figure 2.1 An example of a commercial static microisolator cage for housing mice.

TABLE 2.2: MOUSE CAGE SIZE RECOMMENDATIONS

	Mouse Weight (g)	Cage Floor Area (inch²)	Cage Height (inch)
Mice in groups	<10	6	5
	15	8	5
	25	12	5
	>25	≥15	5
Female + litter		51	5

Source: *The Guide for the Care and Use of Laboratory Animals* (National Research Council 2011).

complete bedding change, though the impact of such an approach on the microbial burden on cage surfaces should be confirmed (Smith et al. 2018).

Mouse cages come in many sizes, and no one size is preferable over any other. However, the cage must be large enough to comfortably accommodate the number of mice that will be housed within. Table 2.2 provides specific recommendations regarding the amount of space required by mice of different sizes and reproductive statuses.

Housing Systems

The following types of housing systems can be used to maintain mice:

1. **Open (no filter top) cages on static (nonventilated) racks**: This type of housing is typically used for conventional mouse colonies (see the section titled "barriers and containment"). It offers no protection against the cage-to-cage spread of undesirable microorganisms. This is the least labor-intensive system.

2. **Filter-top cages on static racks**: This system offers protection against the spread of undesirable microorganisms between cages but is typically associated with the poorest ventilation and offers the greatest potential for a poor microenvironment (Lipman 1999; Perkins and Lipman 1996). Optimum protection against the spread of microorganisms with this method is achieved using ventilated changing stations and appropriate aseptic procedures or *microisolator technique* (National Research Council 2011; White 2007).

3. **Filter-top cages on a rack designed to ventilate each cage individually (i.e., IVC system)**: This system offers good

protection against the spread of undesirable microorganisms, along with the best intracage ventilation of any system (Figures 2.2 and 2.3) (Lipman 1999; 2007). Because the housing of mice on an IVC system can induce chronic cold stress, it is helpful to provide them with materials, such as abundant

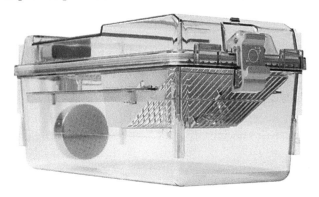

Figure 2.2 A commercially available IVC system cage (Sentry SPP cage; photo courtesy of Allentown, Inc.).

Figure 2.3 A commercially available rack system designed for housing mice in pressurized, individually ventilated cages (Sentry SPP system; photo courtesy of Allentown, Inc.).

bedding or shelters, to allow thermoregulation (David 2013). Optimum protection against the spread of microorganisms is achieved using ventilated changing stations and appropriate aseptic techniques.

4. **Cages (generally open) in a flexible-film or semirigid isolator**: This system offers the best protection against contamination with undesirable environmental microorganisms, with ventilation at least comparable to that achieved using open cages on static racks. Isolators are also recommended when working with zoonotic disease-causing organisms or infectious organisms of mice. For maximum effectiveness, all manipulations, including cage changing, should be performed within the isolator, all materials entering the isolator should be sterilized or decontaminated with a high-level disinfectant, and the isolator port should be thoroughly decontaminated with a high-level disinfectant each time something is introduced into, or removed from, the isolator. This is the most labor-intensive system.

Bedding

Bedding may be used within solid-bottom mouse cages (*contact bedding*) or in catch pans beneath wire-bottom cages (*noncontact bedding*). Unless specifically contraindicated, bedded solid-bottom caging is preferred for housing mice. The bedding absorbs moisture, may enhance reproductive performance, provides some degree of environmental enrichment for the mice, and can significantly reduce odors and microenvironmental contaminants (e.g., ammonia) within cages. Provided that the bedding is sufficiently deep that the mice are able to burrow in it, it also provides at least some degree of thermal insulation for the animals (Freymann et al. 2017; Gordon et al. 1998).

Despite its advantages, bedding can also cause problems if not properly handled. Bedding should be free of splinters and excessive dust. It should also be free of microbial contaminants and insects. Many commercially available bedding materials are sterilized or decontaminated prior to shipping, but recontamination may occur during shipping or storage. Once received at the facility, bedding should be stored off the ground in a vermin-proof room. If possible, it is desirable to sterilize or resterilize it in the facility prior to use. Once opened, it should be stored in a vermin-proof container with a tight-fitting lid. Bedding can also have contaminants that may adversely affect animal or human health, e.g., microbial organisms, endotoxins, or heavy metals (Kaliste et al. 2004; Whiteside et al. 2010). Specifications regarding microbial and inorganic contaminants should be obtained from the manufacturer.

There are many types of contact bedding available for use in mouse cages, including the following:

1. **Wood shavings, shreds, chips, or pellets**: These are the least expensive bedding materials and most (especially shavings) are good for nest building, but they tend to be dusty and offer comparatively poor absorbency and ammonia control. Volatilized hydrocarbons from softwood beddings, particularly cedar or pine, may interfere with experimental results by affecting the activity of liver enzymes (Davey et al. 2003; Vesell 1967). Hardwoods are much less potent in this regard, but hardwood bedding was observed to affect intestinal immune responses in mice (Sanford et al. 2002). The use of wood shavings may facilitate improved nest building and increased litter sizes compared with wood chip bedding (Jackson et al. 2015).

2. **Ground corncob**: These products are typically ground into either ⅛-inch or ¼-inch pieces; they produce little dust and offer fairly good ammonia control but are only moderately absorbent. Corncob can be abrasive, and mold is more of a concern with corncob than with other bedding materials. Corncob is moderately expensive compared with other bedding materials. Mice may also experience comfort issues with cob bedding.

3. **Cellulose, virgin or recycled, shredded or pelleted**: These products vary with regard to absorbency and ammonia control, though the compressed paper has been demonstrated to be more absorbent than corncob bedding (Pallas et al. 2020). They are dusty and more expensive than other types of bedding materials. The shredded products, in particular, tend to be well utilized by mice for nesting.

4. **Pads made of cotton fibers or cotton-based cellulose**: Many, but not all, mice will shred these pads to form a fluffy, highly absorbent bedding that offers excellent ammonia control. The effectiveness of the pads as bedding material is poor

if the mice do not shred them. Bedding pads can be particularly useful for mice with mobility problems, such as EAE or ALS models. Bedding pads are generally expensive.

These products may be blended to improve performance. There are also supplemental materials that can be used to increase absorbency (e.g., ALPHA-dri®, also used as bedding in its own right) or improve reproductive performance by promoting nest building (e.g., Nestlets®, Enviro-Dri™) (Figure 2.4). The ultra-high-absorbency products have been advocated as a primary bedding material for mice that excrete large amounts of urine but must be used with caution as they can be excessively dehydrating, especially for young pups.

Calorie-restricted mice may supplement their diets via consumption of bedding, and this may impact metabolic parameters of research interest (Gregor et al. 2020). Because the type of bedding can influence the degree to which it is consumed, consideration should be given to the role of bedding type as a potential experimental variable in studies that involve calorie restriction.

Mice of both sexes naturally build nests when given appropriate materials to do so. Typical materials include abundant wood shaving or cellulose-based bedding materials, as described above. The

Figure 2.4 Cages bedded with aspen wood chips to which are added crinkle-cut paper, ALPHA-Dri paper pellets, and a Nestlet (left) and crinkle-cut paper and a Nestlet (right).

nest serves several purposes, including helping to protect from hypothermia, serving as a location for the rearing of young, and as protection from intense light. Adults and offspring share the nest, and both parents will retrieve offspring into it. Nest building behavior can be a useful tool in determining the well-being of mice (Gaskill et al. 2013; Rock et al. 2014).

barriers and containment

Laboratory mice may be maintained in a manner that does (*barrier*) or does not (*conventional*) employ specific features and procedures designed to protect them from the introduction and/or dissemination of undesirable microorganisms (National Research Council 1991). The terms *barrier* and *conventional* imply nothing per se about the health status of the mice, only the provisions in place to prevent the introduction and/or spread of contaminants. When barriers are set up to protect mice within the barriers from exposure to undesirable microorganisms, they are viewed as *biosecurity* barriers. When they are used to prevent the spread outside the barriers of undesirable microorganisms harbored by mice within the barriers, they are referred to as *biocontainment* barriers. Both biosecurity and biocontainment barriers combine design features, environmental controls, and operating procedures necessary to achieve the desired goal. These elements are essentially the same for both types of barriers but often function in reverse for biosecurity vs. biocontainment (e.g., shower into a barrier, shower out of biocontainment). In many animal facilities, barriers are designed to provide both biosecurity and biocontainment.

A barrier facility may consist of a single room, a suite of rooms, or an entire building. Barriers may also exist at the level of an isolator holding several cages or even at the level of an individual cage. Individually ventilated racking systems and single microisolator cages can function as effective barriers if animals are handled in ventilated changing stations or biosafety cabinets using appropriate techniques. Barrier facilities vary from maximum to minimum depending on design; methods used to introduce mice, humans, and supplies into the environment; and operational procedures. In the case of maximum barriers used to house very high-health status mice, humans must shower (wet or air shower) and change into sterilized clothing prior to entry, and all supplies—including caging, bedding, and food—must be sterilized. Animals are generally either rederived into the barrier or are of a verified defined-flora health status. Handling of animals is often done using disinfected forceps (Figures 2.5 and 2.6) and may be done within a ventilated animal handling/transfer station (Figure 2.7). Humans need not shower into a minimum barrier, but, if the facility is intended to protect the microbial status of the inhabitants, supplies should be sterilized, disinfected, or sanitized prior to entry. Mice entering such facilities may be of lower health status than those entering higher barriers, but their health status is typically both defined and verified prior to entry. The mice may be handled with forceps or gloved hands.

When properly maintained and operated, isolators provide superior biosecurity and biocontainment. Isolators are available in rigid, semi-rigid, and flexible-film designs. All supplies entering the isolator should be sterilized, and the isolator port should be fogged with a high-level disinfectant (e.g., chlorine dioxide) whenever the port is opened for the introduction or removal of materials or animals. Humans manipulate animals within the isolator from the outside using sleeves with cuffed gloves that are attached to the side of the isolator (Figure 2.8).

For cage-level barriers or facilities that are intended to serve as both biosecurity and biocontainment barriers, security is greatly improved by adhering to a good microisolator technique. Microisolator technique involves opening cages and handling mice only within

Figure 2.5 Picking up a mouse by the tail using forceps.

Figure 2.6 Technique for picking up a newborn mouse using forceps.

Figure 2.7 Ventilated animal handling/transfer station minimizes exposure of animals to contaminants while being handled outside the cage, such as during transfer to clean cages or for experimental manipulation (Photo courtesy of Allentown, Inc.).

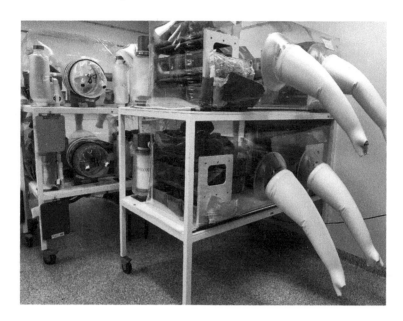

Figure 2.8 Semirigid isolators designed to provide optimum biose-curity and biocontainment for housing mice (Photo courtesy of Betty Theriault, University of Chicago).

ventilated workstations; requiring personnel to wear disposable gloves that are disinfected frequently and/or changed after handling cages of mice; and wiping equipment, including the outside of occupied cages, with a chemical disinfectant prior to entering the workstation and between cages of mice. Additional apparel, such as disposable gowns, disposable sleeve protectors, face masks, and hair bonnets, all serve to further increase the security of the barrier. Personnel who have work assignments in different barriers should always start their day in the barrier housing the highest-health status mice and end their day in the barrier housing the lowest-health status mice. When circumstances dictate working with higher-health status mice after handling contaminated or potentially contaminated mice, a clothes change with or without a shower helps minimize the poten-tial for cross contamination.

personal protective equipment (PPE)

PPE is used within animal facilities for two purposes: to protect the employee and to protect the animals. Barrier clothing, lab coats and gowns, gloves, masks, and bonnets serve both purposes. They limit exposure of the employee to animal allergens and microorganisms carried by the mice, minimize the potential for exposing people outside the facility to these contaminants, and reduce the potential of exposing the mice to microorganisms harbored by humans. Provided that these items are changed between barriers, they can also help prevent the spread of undesirable microorganisms between barriers. Fit-tested disposable respirators or powered air-purifying respirators (PAPR) can be used in place of dust or surgical masks to further reduce exposure of employees to mouse allergens or as prescribed by a physician for allergic employees.

Shower facilities may be incorporated into the barrier entry and/or exit procedure to reduce contaminants carried into or out of the facility by employees. With wet showers, employees shed street clothes prior to showering and don barrier clothing after showering. Air showers, which are designed to remove surface particles from clothing, are most effective with smooth, tightly woven synthetic fabrics. Thus, a change into clean room–type garb or use of an appropriate clean room–type coverall or gown is recommended for optimum effectiveness of this equipment. Proper use of the air shower, including exposure time of >20 s and slow rotation with arms held up, is also important to maximize effectiveness.

special considerations for immunodeficient mice

Immunodeficient mice are less able or unable to eliminate many infectious agents that are readily eliminated by mice with normal immune systems. They are also at greater risk of illness, or even death, from infectious agents that seldom or never cause problems in mice with normal immune systems. As a result, barrier housing is required for the successful maintenance of these animals. The best results will be achieved with isolators, maximum barrier facilities, or microisolator housing in conjunction with a good microisolator technique. Sterilization of cages, including any enrichment devices or other objects that will be placed in the cages, and bedding is recommended. Food for immunodeficient mice should be pasteurized at least, and steam sterilization or irradiation is better whenever possible. Diets that will be subjected to autoclaving must be specially formulated to withstand this treatment, as steam sterilization will alter the concentrations of some vitamins and other nutrients. Drinking water may be a source of microbial contaminants and biofilms that may pose more of a threat to mice with severe immune deficiencies. Acidification (pH 2.5–3.0) or chlorination (6–12 ppm chlorine) will help control some bacterial contaminants, such as *Pseudomonas*. Control of a wider range of infectious agents can be achieved with ultraviolet irradiation or treatment with ozone. Both microbial and nonmicrobial contaminants can be reduced by mechanical filtration or reverse osmosis (Small and Deitrich 2007; Tobin et al. 2007).

environment

The environment in the room (macroenvironment) typically differs from the environment in the cage (microenvironment), which will vary depending on the type of cage, caging system, position on the rack, number of mice, type of bedding, and frequency of cage changing, as well as macroenvironmental conditions. Although it is possible to measure the microenvironment, this is seldom done except for experimental purposes or to investigate specific problems that may be related to environmental conditions. When we talk about controlling and measuring environmental parameters, we are referring to the macroenvironment.

Temperature and Humidity

Environmental temperature and humidity work together in affecting the ability of a mouse to maintain a normal body temperature.

There is no specific environmental temperature or range of temperatures that can be stated unequivocally to promote optimal comfort, health, and performance in laboratory mice. Considerable experience indicates that mice remain healthy and perform well if maintained within a range of 72°F–79°F (22–26°C) (National Research Council 2011). In general, mice prefer temperatures at the upper part of this range or even higher, especially when resting or sleeping (Gaskill et al. 2009; Hankenson et al. 2018). Individually housed and aged animals prefer warmer temperatures than younger, group-housed mice (Gordon 2004). Burrowing, nest building, and huddling are behaviors utilized by mice to maintain body temperature in cooler conditions (Gaskill et al. 2011; Hankenson et al. 2018; National Research Council 2011). Whatever the baseline temperature, it is important to minimize variability in temperature. In this regard, it might be noted that intracage temperature can vary slightly with animal density, and thus it could be important to maintain stable density for studies that require strictly controlled environmental temperature (Toth et al. 2015).

It is generally recommended that rooms housing mice be kept between 30% and 70% relative humidity (National Research Council 2011). There is anecdotal evidence that mice perform best when maintained in the range of 45–60% relative humidity; this appears to be particularly true of wild and wild-derived mice. If the humidity is maintained at this high level, the environmental temperature may be set a bit lower than would be optimal with lower relative humidity.

Ventilation

Ventilation in the room should be adequate to maintain a reasonable concentration of oxygen, minimize levels of gaseous contaminants such as carbon dioxide and ammonia, and dissipate heat generated by mice and equipment. Ventilation rates of 10–15 air changes per hour are recommended for rooms housing mice (National Research Council 2011), though lower room ventilation rates may be acceptable when direct-exhaust IVC systems are used (Geertsema and Lindsell 2015). The air should be fresh, filtered, or both to remove contaminants. Individually ventilated caging can be used to improve ventilation in the mouse's immediate environment.

Illumination

Light intensity in the room during working hours must be sufficient for humans to operate safely and inspect the mice, but not too intense, as higher levels may cause retinal damage in albino animals (La Vail et al. 1987). Light levels not exceeding 325 lux (30 foot-candles) measured at 1 m above the floor have been recommended for rooms housing albino animals (De Vera Mudry et al. 2013; National Research Council 2011).

In many species (e.g., hamster), photoperiod (the intervals of light and dark during a 24-h period) is more important than light intensity and has a critical effect on reproduction. Reproduction in domestic mice is not so dependent on photoperiod, although the light–dark cycle can affect other physiological functions (Nelson and Shiber 1990; Peirson et al. 2018; Petterborg et al. 1984). The two most commonly recommended photoperiods for laboratory mice are 12:12 and 14:10 (light to dark). The latter is typically chosen for breeding colonies.

Shifts in photoperiod and interruptions in the light–dark cycle (e.g., turning on the lights briefly during the dark cycle) can have disruptive effects on mice (Sakellaris et al. 1975), and so these should be avoided. Polycarbonate cages can be either clear or tinted red, blue, or amber. Though tinting is assumed to reduce the intensity of light to which the animals are exposed, it may also result in changes in spectral transmittance that can have physiological impacts on the animals, particularly with respect to melatonin (David et al. 2013). Light intensity can vary markedly within an animal room and depend on factors such as distance from the light source or inadvertent shielding by equipment. Because of this, consideration should be given to practices that minimize the impact of light intensity as a potential experimental variable (Suckow et al. 2017).

Noise and Vibration

Animal rooms are sometimes noisy, and, if excessive, the noise can adversely affect both humans and mice. Noise levels greater than 85 dB are considered potentially damaging to both humans and animals (National Research Council 2011; Perkins and Lipman 1996), with 70 dB suggested as an intracage threshold (Turner 2020), and uncontrolled noise may have adverse effects on mice or their performance in research (Rasmussen et al. 2009; Turner et al. 2007; Willott 2007). To the greatest extent possible, noise levels within the mouse room should be kept well below this limit. This can be achieved with the use of soundproofing materials in room construction and by the separation of mouse rooms from rooms housing noisy species (e.g., dogs or pigs) or noisy activities (e.g., cage washing). It is also important to minimize—preferably eliminate—sudden loud noises from devices such as fire alarms and intercom systems. This is a particular problem with some mouse strains, such as DBA, which are prone to sound-induced seizures. It is important to keep in mind that there is only a partial overlap in the sound frequencies perceived by mice vs. humans, with mice able to detect, and potentially be adversely affected by, ultrasonic noises that are inaudible to humans (Heffner and Heffner 2007). There are many sources of ultrasonic noise in an animal facility, some but not all of which may also emit sounds within the audible range of humans (Sales et al. 1988).

Some facilities advocate the use of "white noise" (e.g., music played at low volume) to limit the disturbing impact on the mice of other sounds associated with work in the animal room, although the *Guide for the Care and Use of Laboratory Animals* advises that "radios, alarms, and other sound generators should not be used in animal rooms unless they are part of an approved protocol or enrichment program" (National Research Council 2011). The *Guide* further advises that any sound generators that are used in animal rooms should be turned off at the end of the day.

Vibration is another variable that is encountered in animal facilities and that may have adverse effects on animals and/or serve as a source of experimental variability (Garner et al. 2018; Turner 2020). Excessive vibration is associated with decreased reproductive success of mice (Atanasov et al. 2015). Vibrations may originate from within the animal room, e.g., from ventilated housing systems or movement of supplies, or outside the facility, e.g., nearby traffic, construction, or trains. Steps should be taken to eliminate unnecessary sources of vibration and minimize or dampen vibrations from sources that cannot be eliminated. An intracage vibration level of no greater than 0.025 g should be maintained (Turner 2020).

sanitation and pest control

Sanitation

There are four components of an effective sanitation program: cage cleaning, room cleaning, equipment cleaning, and quality control.

Cage cleaning

The frequency of cage cleaning will depend on many factors, including the number of mice in the cage, the reproductive status of the mice, the size of the cage, the type of bedding, and cage ventilation (e.g., microisolator cage vs. individually ventilated). It may also vary with the strain of mouse (e.g., diabetic mice produce excessive amounts of urine and must be changed more frequently) and the microbiological status of the mice (e.g., mice that harbor urease-producing bacteria such as *Proteus* will need to be changed more frequently than those that do not, as these bacteria will increase the ammonia concentration in the cage).

As a general guideline, bedding should be changed with sufficient frequency to keep the mice clean and dry and to prevent excessive buildup of ammonia (>25 ppm) in the cage environment. However, this is not a case where more is better. The homecage environment is comforting to mice, and return to the homecage from a foreign environment leads to biochemical changes associated with a rewarding stimulus (Mayer et al. 2021). If bedding is changed too frequently, it can cause undue stress and interfere with breeding. For example, cage changing has been shown to result in physiological changes that can last up to 4 days in some strains of mice (Lim et al. 2019). Also, some mice will kill or neglect their newborn pups if the cage is disturbed too soon after parturition. Mouse cages are often changed concurrently with bedding changes, i.e., the mice are moved to a clean cage with fresh bedding. If this is not done, it is recommended that cages be changed at least once a week (National Research Council 2011). Under certain circumstances, including individually ventilated cages, this period may be extended to 2 weeks or more (National Research Council 2011). Extending the time between cage changes may be possible, though maintenance of acceptable environmental conditions should be validated (Taylor et al. 2019). For some cage components, such as microisolation cage tops, sanitation intervals have been significantly extended based upon the evaluation of microbial load (Ball et al. 2018; Esvelt et al. 2019; Särén et al. 2016).

Mice should be gently handled when transferred to a clean cage. This can be accomplished by means of grasping the mouse by the tail using a gloved hand or forceps (as described earlier) or by scooping them into a plastic transfer cup (Doerning et al. 2019). Nests and young mice can be scooped up by hand and transferred directly to a clean cage. Cupped hand or tunnel handling has been shown to decrease stress associated with handling in mice (Gouveia and Hurst 2017; Hurst and West 2010).

The most effective sanitization of mouse cages and accessories is achieved using an automatic cage washer, although hand cleaning can be effective if done properly. Detergents, chemical disinfectants, hot water, or all of these can be used in this process. Procedures should be adequate to kill vegetative forms of common bacterial contaminants (National Research Council 2011). If chemicals are used, they must be completely removed during the rinsing process. Autoclaving is recommended for caging and equipment that are to be used for immunodeficient mice (see section titled "Special Considerations for Immunodeficient Mice"). As well, cages and wastes from animals that might harbor infectious agents hazardous to personnel or to other animals should be autoclaved before disposal. Treatment with hydrogen peroxide vapor can also be used for microbial decontamination either before or after the use of equipment (Benga et al. 2017).

Chlorine dioxide gas can also be used to decontaminate equipment (Mitchell et al. 2019).

It is important not to neglect the cage rack. Portable racks should be sent through the cage washer on a regular schedule. Fixed racks that cannot be removed from the animal room should be cleaned and disinfected in place. Manifolds for automatic watering systems also need regular attention. In most cases, they are flushed in the wash area at the time of cage cleaning with either large quantities of water or a mild disinfectant followed by water.

Room cleaning

The mouse room should be cleaned daily to remove gross debris. Floors, counters, and so on should also be cleaned daily with a mild disinfectant solution. Walls, ceilings, light covers, ventilation covers, etc. can be cleaned on a less frequent, weekly or monthly schedule. Cleaning utensils such as brooms, mops, and dustpans should be kept in designated areas and not moved from one room to another, as this can be a means of spreading contaminants. More thorough cleaning with a high-level disinfectant or sterilant, such as vaporized hydrogen peroxide, can be done whenever all animals have been, or can be, removed from the room (Dell'Anna et al. 2020; Miedel et al. 2018). To prevent spread of microbial contaminants, thorough cleaning should be done before a new group of mice is moved into a previously occupied room. Chemical sterilants and high-level disinfectants can be harmful, and so precautions must be taken to prevent exposure of humans or other animals.

Research equipment cleaning

Infectious disease may be spread between animals indirectly through research equipment that is used on multiple mice or groups of mice. For example, equipment used for behavioral phenotyping or for imaging is often used by multiple investigators, and mice of different health status may have contact with the equipment. For this reason, it is important to disinfect such equipment between mice or groups of mice. Limited data show that there is no observable effect of cleaning on behavioral assays, such as the elevated plus maze assay (Hershey et al. 2018).

Quality control

The effectiveness of the sanitation program should be monitored on a regular basis. This can be done in part by visual and olfactory inspections for grossly visible dirt and odors. However, these inspections should be supplemented by more objective assessments. For example, the temperatures achieved within the cage washer can be monitored using a heat-sensitive tape that changes color when exposed to the high temperatures needed for effective sanitization. Another approach is to culture surfaces for bacteria or to evaluate equipment surfaces for microbial contamination indirectly via assessment for the presence of adenosine triphosphate (ATP) following disinfection. Professional advice should be sought on methods of obtaining cultures, appropriate media and conditions for incubation, and interpretation of results.

Pest Control

Elimination of insects, feral rodents, and other pests from the vivarium is critical to a quality animal care program. These intruders often harbor disease-causing organisms and may spread contaminants throughout a facility. Poisons designed to eliminate pests pose a potential threat to laboratory mice and should be considered only as a last resort. Humane traps that capture animals alive and that are placed in protected locations near entryways and along walls—including feed and bedding storage areas—are useful for controlling wild and escaped mice (Figure 2.9). Baiting with feed pellets, peanut butter, or sunflower seeds may improve their effectiveness. For both humane and sanitary reasons, traps should be checked at least once daily. Typically, trapped animals are euthanized. Sticky traps are a less humane alternative for feral rodents and should not be used, but they are a desirable method for controlling crawling insects. If these approaches prove inadequate to control a vermin problem, a pest control expert should be consulted. The facility veterinarian should be involved in making any decisions regarding the use of toxic chemicals.

nutrition

Good nutrition is essential to health and optimal performance in laboratory mice, whether it be growth, reproduction, or simply maintenance within the vivarium. Nutrition is also quite complex, as nutrient requirements vary with age, strain, health status, reproductive status, and use in research. It is advisable to purchase a

Figure 2.9 A live trap for escaped or wild mice. Note that food pellets have been placed inside the trap to ensure humane conditions for trapped animals. Traps should be checked daily for trapped animals.

palatable, high-quality mouse food from a reputable supplier of laboratory animal diets. For a review of dietary requirements of laboratory mice, refer to the 1995 publication of the National Research Council's Committee on Animal Nutrition (National Research Council 1995). For specialized needs, consult a veterinarian or nutritionist.

Mice are typically fed *ad libitum* from food hoppers unless food restriction is required as part of the research protocol. It should be noted that, since they are nocturnal, mice consume most of their feed at night (Jensen et al. 2013). The food is generally offered in the form of hard pellets. The hardness of the pellets is important. If the diet is too soft, it will tend to crumble easily, and much will be wasted as the crumbs fall to the bottom of the cage. Soft food will also contribute to an increased incidence of malocclusion. If the food is too hard, the mice may not be able to chew it. Sick mice and mice of some fragile strains may find any pelleted feed too difficult to chew. These mice can be offered ground or moistened feed, which should be changed daily. Some mice will chew or grind food pellets without ingesting much of the feed material, thus leading to significant waste. This behavior may be reduced, in some cases, by providing mice with items such as sunflower seeds.

The feed should be as fresh as possible and never used beyond the manufacturer's recommendations. It also must be stored properly. Ideally, the feed should be stored in a cool, dry, climate-controlled area. Warmth and humidity will hasten deterioration and spoilage. Feed should be kept off the ground and at least a few inches away from walls (Figure 2.10). These precautions will facilitate vermin control, which is essential if the feed is to remain uncontaminated. Once opened, any feed not used immediately should be stored in a vermin-proof container with a tight-fitting lid.

Even feed from a reliable supplier may be contaminated with undesirable microorganisms. For this reason, high-risk mice (e.g., immunodeficient) and other mice maintained in barrier facilities should be given autoclaved or irradiated feed. Autoclaving destroys some nutrients, and so autoclavable feeds are specially fortified so that they will remain nutritionally complete after treatment. Irradiated feed can be purchased directly from some suppliers. Consideration should be given to the temperature used to autoclave feed, as the

Figure 2.10 Unopened bags of feed stored on a cart that maintains the feed off the floor. A mobile cart such as this allows for ease of cleaning in the event of spillage.

carcinogen acrylamide can be produced as a result of autoclaving (Kurtz et al. 2018).

water

Mice can be provided with water from water bottles or automatic watering systems. Automatic watering systems offer the advantages of convenience and reduced labor. However, they can cause serious flooding—a particular problem for mice housed in solid-bottom cages—and they can serve as sources of microbial contamination if the water lines are not routinely sanitized. Water bottles are labor intensive but are preferable for situations where bacterial contamination is unacceptable (e.g., severely immunodeficient animals). Though glass water bottles had previously been common, they have largely been displaced by plastic bottles. Of note, bisphenol A, an endocrine-disrupting chemical, can leach into the water from polycarbonate bottles, and thus consideration should be given to the material of which water bottles are made (Honeycutt et al. 2017). Water for high-risk or otherwise protected colonies should be treated to minimize or eliminate contamination. This can be accomplished by a variety of techniques, including autoclaving, ultraviolet irradiation, ultrafiltration, or reverse osmosis. Chlorination (6–12 ppm) or acidification (pH 2.5–3.0) can be used alone or in addition to the aforementioned methods to prevent colonization with *Pseudomonas*, but these treatments will not eliminate established infections. Water quality is often monitored to assess microbiological contaminants. Further, it may be worthwhile to monitor acidified water provided via water bottles for heavy metals that can leach out of water bottle components (Nunamaker et al. 2013).

breeding

Breeding laboratory mice is seldom as simple as putting male and female together and waiting for the babies (called *pups*) to appear. The genetic background of the mice selected for breeding is almost always important, and for many purposes it is critical. For genetic experiments, the criteria for the selection of breeders are quite specific and may vary from one generation to the next. Even for more general purposes, failure to choose breeders of the appropriate genetic background will eventually result in unwanted—albeit often subtle—changes in the characteristics of the mice. The following discussion covers some of the more basic aspects of breeding laboratory mice. More complete examinations of the topic can be found in other sources (Berry and Linder 2007; Currer et al. 2009; Dorsch 2012).

Basic Genetics

Any discussion of mouse breeding requires an understanding of a few basic genetic terms:

- **Gene**: A single unit of DNA that contains the code for a specific trait.
- **Allele**: One form of a specific gene. For most genes, an individual animal carries one allele inherited from its mother and one allele inherited from its father. These alleles may be the same or they may be different. Example: Gene 1 may have alleles a, b, c, d, or e.
- **Wild-type**: The most common allele in a population.
- **Homozygous**: Having the same allele from both parents for the same gene. Example: A specific mouse has two copies of the b allele for Gene 1 (one from its father and one from its mother).
- **Heterozygous**: Having a different allele from each parent for the same gene. Example: A specific mouse has one copy of the b allele for Gene 1 (inherited from its father) and one copy of the c allele (inherited from its mother).
- **Inbred strains**: Mice maintained through pedigreed, brother × sister mating. All mice from a particular inbred strain are homozygous for virtually every gene. If mated together, two inbred mice produce offspring that are genetically identical to each other and to their parents. Example: All mice from inbred strain Q carry two copies of the a allele for Gene 1, two copies of the e allele for Gene 2, two copies of the b allele for Gene 3, etc.
- **Hybrid**: Hybrid mice have a dam of one genetic background and a sire of a different genetic background. F1 (first filial generation) hybrid mice have a dam of one inbred strain and a sire of a different inbred strain. Although they are heterozygous for many different genes, the heterozygosity is limited in that

there are only two alleles for each gene (one from the maternal inbred strain and one from the paternal inbred strain) within the entire population of hybrid animals. F1 hybrid animals are therefore genetically identical to each other, although, if mated together, two F1 hybrids would produce offspring that differed genetically from each other and from their parents. Example: All F1 hybrid mice resulting from a mating between strain Q and strain Z carry the a and b alleles for Gene 1, the e and d alleles for Gene 2, the b and c alleles for Gene 3, etc. F2 hybrid mice result from mating two F1 mice. This results in a reassortment of alleles so that F2 mice are not genetically identical to one another or any of their parents or grandparents.

- **Outbred stocks**: Mice maintained through a breeding system that seeks to maximize heterozygosity and avoid mating related mice. Mice from a particular outbred stock are heterozygous for many genes. Outbreds differ from hybrids in that, within the entire stock, each gene has different alleles that combine in different ways within individual animals. If mated together, two outbred mice would produce offspring that differed genetically from each other and from their parents. Example: One mouse from outbred stock T may carry the e and b alleles for Gene 1, the a and c alleles for Gene 2, the b and c alleles for Gene 3, etc.; another mouse from outbred stock T may carry the c and d alleles for Gene 1, the b and f alleles for Gene 2, the a and e alleles for Gene 3, etc.

Breeding Schemes

Genetic experiments aside, breeding schemes for laboratory mice are generally designed to either:

1. Develop or preserve a particular characteristic or group of characteristics; or
2. Maintain maximum genetic variability.

Maintaining maximum genetic variability is almost always desirable when breeding outbred stocks of mice (e.g., ICR). This is accomplished by random breeding of unrelated individuals. Even with random breeding, however, there will be a loss of genetic variability within a closed colony; this will happen rapidly in a small colony versus more slowly in a large colony. For this reason, maintenance of outbred stocks is generally not recommended for individual researchers.

With inbred mice, including inbred mice carrying spontaneous mutations, the goal is to preserve genetic uniformity in the colony. With most strains, this involves nothing more than mating males and females of the same strain, but strict inbreeding involves only brother × sister mating. The only precaution is to avoid creation of a substrain, which occurs when a closed colony of inbred mice is isolated for at least 20 generations. Generations are counted at both the colony of origin and the separated colony, and so 20 generations are reached surprisingly quickly. A substrain will differ genetically from the original inbred strain and may yield different results when used in research. To avoid this, it is important to introduce outside animals from the same inbred strain on a regular basis, unless fastidious genetic monitoring is performed to ensure lack of genetic drift. Some mutant mice do not make good parents because they are infertile or have poor maternal instinct. More complex breeding schemes are necessary to maintain these mice (Berry and Linder 2007; Currer et al. 2009).

Transgenic and knockout mice may be more difficult to breed than inbred mice. Among the many problems that may be encountered with these animals is a failure of homozygotes to survive to breeding age. If homozygotes are needed for a study, it may be necessary to use an alternative scheme, such as breeding heterozygotes (Berry and Linder 2007; Currer et al. 2009).

Such matings will result in some homozygotes (statistically, about 25% of the pups), some heterozygotes (about 50% of the pups), and some pups that do not carry the mutation at all. As all of the pups may look the same, genetic testing is often required to distinguish those that carry the mutation from those that do not.

Hybrid mice are desirable for some experiments. Breeding hybrids requires mating a male mouse of one inbred strain with a female mouse of another inbred strain. The resulting pups are first-generation, or F1, hybrids. In some cases, second-generation (F2) hybrids are desired. These are obtained by mating an F1 female with an F1 male. Except for genetic experiments, it is rarely desirable to breed hybrids past the second generation. So, to keep the hybrid colony going, it is necessary to maintain colonies of the inbred parental lines to serve as breeders in the hybrid colony.

The maximum number of pups per female mouse is obtained by leaving the female with an adult male throughout her breeding life.

Usually, mice become sexually mature by 7–8 weeks of age. Mice are continuously polyestrous and have an estrous cycle that typically lasts 4–5 days. The breeding productivity of the male can be increased using a harem system (two or more females per male). Trio breeding is a common approach and describes the situation in which one male and two females are housed together, usually continuously. Trio breeding allows for the dams to assist in raising each other's pups, and there are few marked differences in reproductive success between pair-bred and trio-bred mice (Wasson 2017). Although mice can reproduce well beyond 1 year of age, their reproductive performance will typically diminish after 8–10 months of age. Other breeding systems can be used to achieve different goals. For example, if the goal is to maximize the number of pups weaned per litter, this is best achieved by moving the female to a separate cage prior to parturition (e.g., when she is visibly pregnant). If the goal is to maximize the genetic contribution to the colony from a single male, this is best achieved by mating him with many females using a rotating system, in which the male is moved between cages of females. Both of these systems will reduce female productivity, however, as no male will be present to breed with the female during her postpartum estrus, which occurs within hours after the birth of the litter. The absence of a male in a cage may also affect the development of female pups. The next opportunity for the female to become pregnant will be after the litter is weaned.

Neonatal mortality can be a significant issue in some breeding colonies. Although mortality of pups can be related to the phenotype of genetic mutants, it can also involve a number of other factors. In this regard, maternal care of the pups should be evaluated. Close observation of dams and litters may help determine if the dam has no milk or is not feeding the pups. For example, one should evaluate whether or not the pups move normally and attempt to nurse. Poor mothering may be influenced by strain, environment, and the pup itself. If litters are very large, weaker pups may not be able to nurse sufficiently. Removal of some pups or cross-fostering to other dams with similarly aged litters may promote survival of weaker pups. For some mouse strains, provision of environmental enrichment can help reduce pup mortality (Leidinger et al. 2019). Busy, loud environments are generally not well tolerated, and some strains and dams seem to be especially sensitive to such conditions. Mice being bred may benefit from transfer to a quiet environment, no disturbance of the crate for a week after pups are first noted in the cage, and

TABLE 2.3: TYPICAL REPRODUCTIVE PARAMETERS OF THE MOUSE

Parameter	Typical Value
Age of sexual maturity	7–8 weeks
Estrous cycle length	4–5 days
Gestation length	19–21 days
Litter size	10–12 pups
Birth weight	1 g
Age at weaning	21–28 days
Pseudopregnancy	10–13 days

provision of copious amounts of nesting material. Breeding success may be improved when two females, particularly littermate females, are housed together and can assist each other in caring for the pups.

Under some conditions, female mice have prolonged estrous cycles or become pseudopregnant. The prolongation (or cessation) of cycles in group-housed female mice is known as the Lee–Boot effect. Pseudopregnancy can result from coitus or from stimulation of the vagina and cervix by swabbing for vaginal cytology. Pseudopregnant mice may exhibit behavior similar to a pregnant mouse; however, these changes begin to diminish after 10–13 days of pseudopregnancy.

Although reproductive parameters will vary with the strain/stock and age of the mouse and environmental conditions, typical values are presented in Table 2.3.

Embryo transfer is sometimes used as a way to produce mice following genetic modification of mouse embryos. Once the genetic manipulation has been done, the embryos are transferred surgically to the oviduct or uterine horn of a pseudopregnant mouse. Methods for the nonsurgical transfer of embryos to recipient mice allow this to be achieved in a way that eliminates the pain associated with surgery (Cui et al. 2014; Steele et al. 2013). Because it is possible to cryopreserve mouse embryos and even sperm, valuable lines of mice can be banked and reanimated for use in embryo transfer, and this allows archiving and disaster recovery of important genotypes (Dorsch 2012; Takeo et al. 2020).

Pheromone Influences

Pheromones are chemicals produced by animals of many species that evoke specific stereotypical responses in other animals of the same species. In mammals, including mice, pheromones appear to be detected primarily by the vomeronasal organ, a sensory organ that is located at the base of the nasal septum, although the main olfactory system also plays a role in pheromone detection. Pheromones produced by mice can significantly influence reproductive physiology and various social behaviors, including sexual behavior. For example, exposure of immature female mice to urine from mature males (Bronson and Maruniak 1975) or females in estrus (Drickamer 1982) accelerates the onset of puberty, whereas urine from group-housed females delays the onset of puberty in immature females (Bronson and Maruniak 1975). Estrous cycling tends to cease in group-housed females isolated from males (Lee–Boot effect), but cycling resumes when the noncycling females are exposed to an adult male or his urine (Whitten effect). The Lee–Boot and Whitten effects can be used to synchronize estrus and facilitate timed pregnancy; estrus generally occurs 3 days after first exposure of noncycling females to a male or male urine. Pregnancy disruption (implantation failure) and pseudopregnancy may be induced by exposure of a pregnant female to urine from an unfamiliar male within a few days of insemination (Bruce effect). Both pheromone production and pheromone-induced behaviors can be influenced by other factors, e.g., housing density (Coppola and Vandenbergh 1985), food deprivation (Drickamer 1984a), photoperiod (Drickamer 1984a), and season (Drickamer 1984b).

Timed Pregnancy

For some experimental procedures, it is necessary to know the date of conception of a litter (e.g., when prenatal mice at a certain day of gestation are needed). This is done by placing the female with a male and then checking her early the following morning for the presence of a vaginal plug. This is a whitish mass composed of coagulated secretions from the coagulating and vesicular glands of the male mouse, and it typically fills the vagina of the female for 8–24 h following breeding (which generally occurs during the dark part of the light cycle). To visualize the plug, one should lift the female by the base of her tail and, if necessary, spread the lips of the vulva slightly with forceps or flat-tipped toothpick. If there is no plug, the female is left with the male and checked each morning until a plug is found. While the presence of a plug is no guarantee of pregnancy, it indicates that sexual activity occurred, and pregnancy is likely. The first day of gestation is considered to be the day after the plug is found.

The Whitten effect can be exploited to synchronize estrus when larger numbers of timed pregnant mice are needed. When noncycling females are placed with males (typically one–three females per male), the estrous cycle will restart, and most will enter estrus within 3 days. Daily examination of the females for vaginal plugs will identify those that have mated and are likely to be pregnant.

Genetic Monitoring

Genetically defined mice are valued in research precisely because they are genetically defined. When a researcher selects a particular strain for an experiment, they count on the fact that mice of this strain will reliably exhibit certain characteristics, and that these mice will respond to experimental manipulations in the same manner as other mice of the same strain used in previous experiments and in other laboratories. It is therefore essential that the investigator knows with certainty that their C57BL/6 mice are indeed C57BL/6 and not some other strain of black mice. Furthermore, the researcher must be confident that a black mouse of another strain has not escaped and mated with one of their C57BL/6 mice, resulting in the production of hybrids that appear identical to the C57BL/6 mice. As substrains of the same strain—for example, C57BL/6J vs. C57BL/6N—also differ from each other in some respects, it is important to know which substrain is being used.

In many cases, it is easy to detect an impostor. For example, no one will mistake a BALB/c mouse for a C57BL/6 mouse because the BALB/c is albino and the C57BL/6 is black. Similarly, it will be easy to tell if a BALB/c has gotten out and mated with a C57BL/6. The offspring will be agouti (where the shaft of the hair is one color—usually yellow or red—and the tip is black, giving the animal a golden color), not black. However, a black C57BL/10 mouse cannot be distinguished visually from a C57BL/6 mouse, and an unplanned mating between these two strains will produce pups that are indistinguishable from either parent. For this reason, it is best, whenever possible, to keep strains of the same coat color in separate rooms, cubicles, etc. It is also essential to identify individual mice and keep accurate records, including properly labeled cage cards with correct strain nomenclature. If a mouse escapes, it should not be returned to the colony unless its identity can be determined with absolute certainty.

In addition to taking appropriate precautions to guard against inadvertent mix-ups between mice of similar-appearing strains, it may be necessary to assay genetic consistency of the mice. If there is any suspicion that a genetic contamination may have occurred (e.g., that a C57BL/10 mouse may have gotten into the C57BL/6 colony), genetic testing should be performed. Periodic genetic testing is also advisable whenever breeding colonies of mice of the same coat color are maintained in the same room, particularly if the colonies are handled by different research or animal care personnel. Similarly, since mutations do occur and can lead to significant alterations in the genetic makeup of a colony, genetic testing should be performed whenever closed breeding colonies are maintained over a long period of time.

Older methods of genetic testing included electrophoretic testing for biochemical markers, serologic testing for immunological markers, skin grafting, mixed lymphocyte reaction testing, karyotype analyses, osteometric trait analysis, and screening for coat color. These methods have largely been replaced by genomic testing that is based on assays for polymorphic genetic markers (Benavides et al. 2020). In the case of inbred strains, genetic uniformity is the goal; in the case of outbred stocks, genetic heterogeneity is desired. In addition to a program of genetic testing, it is wise to remove from a breeding colony any mouse that is born with an obvious mutation (e.g., white spots on a mouse of a solid black strain). Technicians should be trained to recognize the normal appearance and behavior of strains with which they work.

identification and record keeping

Identification

Proper identification of mice is imperative for effective management of breeding colonies as well as for most research studies. All mice should be identified by a cage card, and, if appropriate, group-housed mice also should be identified individually.

Cage cards are typically placed on individual cages of mice, although, under certain circumstances, it may be acceptable to have a single card for an entire section or room of mice. Cage cards of different colors can be used to facilitate distinction between mice of different strains, mice belonging to different investigators, or mice being used on different research protocols. The following information should be noted on cage cards.

1. **Complete and correct nomenclature**: Use of abbreviated nomenclature can be tempting but will almost inevitably lead to confusion and potentially serious mix-ups.
2. Source of the animals and their age or date of birth.
3. **Sex**: If the cage contains more than one animal, note the number of animals of each sex.
4. Name of the responsible investigator and, if appropriate, the protocol number of the research project for which the animals will be used.
5. In breeding colonies, the cage card may indicate pedigree number and reproductive performance history (e.g., date mated, birth and weaning dates of litters, and/or the number of pups born and weaned).
6. It is also helpful to note on the cage card or other easily accessible records any procedures performed (e.g., surgery, tumor implantation).

Common methods for permanent identification of individual mice include the following (Dahlborn et al. 2013):

1. **Ear tags**: Tags designed for use on mice are commercially available. A disadvantage is that ear tags can be lost, especially in a cage where mice are fighting.
2. **Ear punch**: For small groups of mice, e.g., a single cage or litter, a simple nonnumeric punch system may be useful. However, for larger groups of animals, this technique is effective only if a consistent numbering system is employed and personnel are trained to use it. Ear punching is considered to be relatively painless for mice (Taitt and Kendall 2019).
3. **Fur marking**: The fur can be marked with non-toxic markers or dyes for temporary identification. It is essential that applied materials are non-toxic and will not affect any experimental parameters if ingested or absorbed by the animals. Fur marking allows for easily applied identification, but it is only temporary, as the markings may fade as the animals groom themselves.
4. **Tattoo**: Methods for tattooing mice, usually on the tail, footpad, ears, or toe, alone or in combination, can be effectively used for identification (Chen et al. 2016). With appropriate

equipment, mice of all ages can be identified in this way, even neonatal mice.

5. **Subcutaneous transponder**: Anesthesia of the mouse prior to insertion of the transponder is recommended. A device designed to detect the transponder must be used as part of this system.

6. **Toe clip**: This method can be used to identify very young (typically, up to 7 days) pups when no other method is suitable, and the sampled material should be used for DNA analysis. Aseptic technique should be used when performing this procedure.

Records

Records are kept within an animal facility for a variety of purposes, including documentation of health monitoring and veterinary care, tracking and census, and colony management. Records may be electronic and/or written paper documents or ledgers. Records related to environmental conditions (e.g., temperature, relative humidity) and sanitation should be maintained for a period sufficient to allow retrospective assessment of any animal health issues or perturbations in research outcomes.

Individual health records are generally not maintained for mice. However, it is important to keep "herd" records, documenting health problems within a colony, and records delineating treatments of individual mice under veterinary care. Accurate census records are essential for determining per diem charges and can prove invaluable for managing and assigning space within a facility. Census records for mice are typically determined by counting cages rather than individual mice. Colony management records are necessary for breeding colonies but may not be needed for research colonies in which there is no breeding. The following types of records are recommended for breeding colonies.

1. **Pedigree charts**: A pedigree chart is essentially a register delineating the ancestors, progeny, and other relatives of a particular mouse. It may be kept electronically or as a handwritten ledger, traditionally in a bound book. Today, there are computer software packages available for this specific purpose. Pedigree charts are necessary to maintain maximum genetic diversity within outbred stocks and to facilitate

elimination of undesirable traits (i.e., spontaneous mutations) within inbred strains. Within large breeding colonies, a
small foundation colony may be maintained as the source for
a larger research or production colony. A foundation colony
is a small, isolated colony that is used to produce breeding
stock for the larger colony, whereas a production colony is one
that is maintained for the sole purpose of producing as many
mice as possible within a given space. In such cases, pedigree
records are maintained for the small foundation colony but
may not be maintained for the larger colony.

2. **Breeding records**: It is ideal to keep a record of the reproductive performance of all breeding animals. This is often accomplished using a cage card, which may be designed specifically
   for the purpose. The reproductive record should indicate the
   date of birth of each litter, the number of pups of each sex
   born, and the number of pups of each sex weaned. Combined
   with a good pedigree system, this information can be used to
   select the most productive animals and lines and eliminate
   poorly productive animals and lines, thereby increasing the
   overall productivity of the colony and minimizing the cost of
   maintaining nonproductive animals.

transportation

Between Institutions

Animal facilities commonly exchange mice with other institutions
around the world. Transportation is inevitably stressful for the animals, but stress can be minimized and the safety of the animals
improved by attention to the following issues.

1. **The shipping container**: The shipping container must be
   durable enough to protect the mice during transit and to
   prevent them from chewing their way out. It must be free of
   sharp edges or other features that may trap or injure the animals. It should be resistant to moisture, and it must be adequately ventilated. Ventilation openings must be covered with
   a chew-resistant screen and should be covered with a filter
   material to protect the mice from pathogenic organisms present in the environment. The box must be of sufficient size to

accommodate all mice in the shipment without undue crowding. Keep in mind, however, that huddling with other mice can greatly reduce the risk of hypothermia during shipment in cold weather. Containers for mice that will be shipped by air should conform to the regulations of the International Air Transport Association (www.iata.org). Shipping containers suitable for transportation of mice can be purchased from many commercial mouse vendors. Many institutions clean the outside of shipping containers with a disinfectant upon entry into the animal facility as a way to minimize the possibility of infectious agent spread.

2. **Feed, water, and bedding**: The shipping container should contain sufficient bedding and nesting material to absorb urine and provide insulation from the cold. A source of moisture should be provided, and food should be provided if the mice will be in transit for more than a few hours. Commercially available water/gel packs can be used as a water source. Alternatives such as wet chow mash or sliced potatoes can provide both moisture and calories.

3. **Health records**: Most facilities receiving mice will want a record of the health status of the animals, and such a record is required for most international shipments. Many countries outside the United States also require a health certificate that may or may not have to be endorsed by a veterinarian employed by the US Department of Agriculture (USDA). Health certification requirements vary significantly from country to country and are subject to change. To avoid having your mice held up in customs, it is imperative to ascertain the exact requirements of the destination country prior to shipment. The health report, with or without a health certificate, should be attached to the outside of the box in such a manner that they are easily accessible during shipment.

4. **Requirements during shipment**: To the greatest extent possible, mice should be kept in a dry, well-ventilated, climate-controlled environment during shipping. Prolonged exposure to extremes of temperature must be avoided. Carrier(s) should be selected, and shipping times arranged, with these requirements in mind. Coordination between the shipping and receiving institutions will also help ensure that any problems that might occur during shipping are quickly detected and rectified.

Within the Institution

It is often necessary, even routine in many institutions, to transport mice from one building or part of a building to another. Many of the requirements for transportation between institutions are not necessary for intra-institutional transportation, but attention still must be paid to the shipping container and temperature and air quality within the container. Efforts also should be made to protect the mice from contamination with unwanted microbial agents, and to protect humans from exposure to mouse allergens.

Mice may be transported in their own cages or in transport containers designed for this purpose. In general, the cages/containers should be bedded, but food and water typically are not necessary unless the animals will be in transit for more than a couple of hours. Water bottles may actually be undesirable, as the jostling during transit may cause them to run out over the mice and bedding. To protect both animals and humans, cages/containers should be covered during transport, particularly if animals will be transported through public areas. Transportation should be planned to minimize transit time and ensure that someone will be available to receive the animals at the final destination; this is especially important if filter tops or covers will impede ventilation within the shipping container. Exposure to adverse weather conditions or extremes of temperature should be avoided or minimized.

---

references

Atanasov NA, Sargent JL, Parmigiani JP, Palme R, Diggs HE. 2015. Characterization of train-induced vibration and its effect on fecal corticosterone metabolites in mice. *J Am Assoc Lab Anim Sci* **54**:737.

Ball BLS, Donovan KM, Clegg S, Sheets JT. 2018. Evaluation of extended sanitation interval for cage top components in individually ventilated mouse cages. *J Am Assoc Lab Anim Sci* **57**:138.

Benavides F, Rülicke T, Prins J-B, Bussell J, Scavizzi F, Cinelli P, Herault Y, Wedekind D. 2020. Genetic quality assurance and genetic monitoring of laboratory rats and mice: FELASA working group report. *Lab Anim* **54**:135.

Benga L, Benten WPM, Engelhardt E, Gougoula C, Schulze-Röbbecke R, Sager M. 2017. Survival of bacteria of laboratory animal origin on cage bedding and inactivation by hydrogen peroxide vapour. *Lab Anim* **51**:412.

Berry ML, Linder CC. 2007. Breeding systems: considerations, genetic fundamentals, genetic background, and strain type. In *The Mouse in Biomedical Research, 2nd ed, Vol I (History, Wild Mice, and Genetics)*, Fox JG, Barthold SW, Davisson MT, Newcomer CE, Quimby FW, Smith AL, Eds. Boston: Academic Press, 53.

Bronson FH, Maruniak JA. 1975. Male-induced puberty in female mice: evidence for a synergistic action of social cues. *Biol Reprod* **13**:94.

Chen M, Kan L, Ledford BT, He J-Q. 2016. Tattooing various combinations of ears, tails, and toes to identify mice reliably and permanently. *J Am Assoc Lab Anim Sci* **55**:189.

Coppola DM, Vandenbergh JG. 1985. Effect of density, duration of grouping and age of urine stimulus on the puberty delay pheromone in female mice. *J Reprod Fertil* **73**:517.

Cui L, Zhang Z, Sun F, Duan X, Wang M, Di K, Li X. 2014. Transcervical embryo transfer in mice. *J Am Assoc Lab Anim Sci* **53**:228.

Currer JM, Corrow D, Strobel M, Flurkey K. 2009. Breeding strategies and techniques. In *The Jackson Laboratory Handbook on Genetically Standardized Mice, 6th ed.*, Flurkey K, Currer JM, Leiter EH, Witham B, Eds. Bar Harbor: The Jackson Laboratory, 241.

Dahlborn K, Bugnon P, Nevalainen T, Raspa M, Verbost P, Spangenberg E. 2013. Report of the Federation of European Laboratory Animal Science Associations Working Group on animal identification. *Lab Anim* **47**:2.

Davey AK, Fawcett JP, Lee SE, Chan KK, Schofield JC. 2003. Decrease in hepatic drug-metabolizing enzyme activities after removal of rats from pine bedding. *Comp Med* **53**:299.

David JM, Knowles S, Lamkin DM, Stout DB. 2013. Individually ventilated cages impose cold stress on laboratory mice: a source of systemic experimental variability. *J Am Assoc Lab Anim Sci* **52**:738.

Dell'Anna G, Mullin K, Brewer MT, Jesudoss Chelladurai JRJ, Sauer MB, Ball BLS. 2020. Room decontamination using ionized hydrogen peroxide fog and mist reduces hatching rates of *Syphacia obvelata* ova. *J Am Assoc Lab Anim Sci* **59**:365.

De Vera Mudry MC, Kronenberg S, Komatsu S, Aguirre GD. 2013. Blinded by the light: retinal phototoxicity in the context of safety studies. *Toxicol Pathol* **41**:813.

Doerning CM, Thurston SE, Villano JS, Kaska CL, Vozheiko TD, Soleimanpour SA, Lofgren JL. 2019. Assessment of mouse handling techniques during cage changing. *J Am Assoc Lab Anim Sci* **58**:767.

Dorsch MM. 2012. Cryopreservation of preimplantation embryos and gametes, and associated methods. In *The Laboratory Mouse, 2nd ed.*, Hedrich HJ, Ed. San Diego: Elsevier, 675.

Drickamer LC. 1982. Delay and acceleration of puberty in female mice by urinary chemosignals from other females. *Dev Psychobiol* **15**:433.

Drickamer LC. 1984a. Urinary chemosignals and puberty in female house mice: effects of photoperiod and food deprivation. *Physiol Behav* **33**:907.

Drickamer LC. 1984b. Seasonal variation in acceleration and delay of sexual maturation in female mice by urinary chemosignals. *J Reprod Fertil* **72**:55.

Esvelt MA, Steiner L, Childs-Thor C, Dysko RC, Villano JS, Freeman ZT. 2019. Variation in bacterial contamination of microisolation cage tops according to rodent species and housing system. *J Am Assoc Lab Anim Sci* **58**:450.

Freymann J, Tsai P-P, Stelzer HD, Mischke R, Hackbarth H. 2017. Impact of bedding volume on physiological and behavioural parameters in laboratory mice. *Lab Anim* **51**:601.

Garner AM, Norton JN, Kinard WL, Kissling GE, Reynolds RP. 2018. Vibration-induced behavioral responses and response threshold in female C57BL/6 mice. *J Am Assoc Lab Anim Sci* **57**:447.

Gaskill BN, Karas AZ, Garner JP, Pritchett-Corning KR. 2013. Nest building as an indicator of health and welfare in laboratory mice. *J Vis Exp* **82**:e51012.

Gaskill BN, Rohr SA, Pajor EA, Lucas JR, Garner JP. 2009. Some like it hot: mouse temperature preferences in laboratory housing. *Appl Anim Behav Sci* **116**:279.

Gaskill BN, Rohr SA, Pajor EA, Lucas JR, Garner JP. 2011. Working with what you've got: changes in thermal preference and behavior in mice with or without nesting material. *J Therm Biol* **36**:193.

Geertsema RS, Lindsell CE. 2015. Effect of room ventilation rates in rodent rooms with direct-exhaust IVC systems. *J Am Assoc Lab Anim Sci* **54**:521.

Gouveia K, Hurst JL. 2017. Optimising reliability of mouse performance in behavioural testing: the major role of non-aversive handling. *Sci Rep* **7**:44999.

Gordon CJ. 2004. Effect of cage bedding on temperature regulation and metabolism of group-housed female mice. *Comp Med* **54**:63.

Gordon CJ, Becker P, Ali JS. 1998. Behavioral thermoregulatory responses of single- and group-housed mice. *Physiol Behav* **65**:255.

Gorence GJ Pulcastro HC, Lawson CA, Gerona RR, Friesen M, Horan TS, Gieske MC, Sartain CV, Hunt PA. 2019. Chemical contaminants from plastic in the animal environment. *J Am Assoc Lab Anim Sci* **58**:190.

Gregor A, Fragner L, Trajanoski S, Li W, Sun X, Weckwerth W, König J, Duszka K. 2020. Cage bedding modifies metabolic and gut microbiota profiles in mouse studies applying dietary restriction. *Sci Rep* **10**:20835.

Hankenson FC, Marx JO, Gordon CJ, David JM. 2018. Effect of rodent thermoregulation on animal models in the research environment. *Comp Med* **68**:425.

Heffner HE, Heffner RS. 2007. Hearing ranges of laboratory animals. *J Am Assoc Lab Anim Sci* **46**:20.

Hershey JD, Gifford JJ, Zizza LJ, Pavlenko DA, Wagner GC, Miller S. 2018. Effects of various cleaning agents on the performance of mice in behavioral assays of anxiety. *J Am Assoc Lab Anim Sci* **57**:335.

Honeycutt JA, Nguyen JQT, Kentner AC Brenhouse HC. 2017. Effects of water bottle materials and filtration on bisphenol A content in laboratory animal drinking water. *J Am Assoc Lab Anim Sci* **56**:269.

Howdeshell KL, Peterman PH, Judy BM, Taylor JA, Orazio CE, Ruhlen RL, Vom Saal FS, Welshons WV. 2003. Bisphenol A is released from used polycarbonate animal cages into water at room temperature. *Environ Health Perspect* **111**:1180.

Hurst JL, West RS. 2010. Taming anxiety in laboratory mice. *Nat Methods* **7**:825.

Jackson E, Demarest K, Eckert WJ, Cates-Gatto C, Nadav T, Cates LN, Howard H, Roberts AJ. 2015. Aspen shaving versus chip bedding: effects on breeding and behavior. *Lab Anim* **49**:46.

Jensen TL, Kiersgaard MK, Sørensen DB, Mikkelsen LF. 2013. Fasting of mice: a review. *Lab Anim* **47**:225.

Kaliste E, Linnainmaa M, Meklin T, Torvinen E, Nevalainen A. 2004. The bedding of laboratory animals as a source of airborne contaminants. *Lab Anim* **38**:25.

Kurtz DM, Glascoe R, Caviness G, Locklear J, Whiteside T, Ward T, Adsit F, Lih F, Deterding LJ, Churchwell MI, Doerge DR, Kissling GE. 2018. Acrylamide production in autoclaved rodent feed. *J Am Assoc Lab Anim Sci* **57**:703.

La Vail MM, Gorrin GM, Repaci M. 1987. Strain differences in sensitivity to light-induced photoreceptor degeneration in albino mice. *Curr Eye Res* **16**:825.

Leidinger CS, Thöne-Reineke C, Baumgart N, Baumgart J. 2019. Environmental enrichment prevents pup mortality in laboratory mice. *Lab Anim* **53**:53.

Lim MA, Defensor EB, Mechanic JA, Shah PP, Jaime EA, Roberts CR, Hutto DL, Schaevitz LR. 2019. Retrospective analysis of the effects of identification procedures and cage changing by using data from automated, continuous monitoring. *J Am Assoc Lab Anim Sci* **58**:126.

Lipman NS. 1999. Isolator rodent caging systems (state of the art): a critical view. *Contemp Top Lab Anim Sci* **38**:9.

Lipman NS. 2007. Design and management of research facilities for mice. In *The Mouse in Biomedical Research, 2nd ed., Vol. III (Normative Biology, Husbandry, and Models)*, Fox JG, Barthold SW, Davisson MT, Newcomer CE, Quimby FW, and Smith AL, Eds. Boston: Academic Press, 271.

Mayer FP, Iwamoto H, Hahn MK, Grumbar GJ, Stewart A, Li Y, Blakely RD. 2021. There's no place like home? Return to the home cage triggers dopamine release in the mouse nucleus accumbens. *Neurochem Int* **142**:104894.

Miedel EL, Ragland NH, Engelman RW. 2018. Facility-wide eradication of *Corynebacterium bovis* by using PCR-validated vaporized hydrogen peroxide. *J Am Assoc Lab Anim Sci* **57**:465.

Mitchell CM, McGrath A, Beck B, Schurr MJ, Fong D, Leszczynski JK, Manuel CA. 2019. Low-cost, small-scale decontamination of laboratory equipment by using chlorine dioxide gas. *J Am Assoc Lab Anim Sci* **58**:569.

National Research Council. 1991. Barrier programs. In *Infectious Diseases of Rats and Mice*. Washington, DC: National Academies Press, 17.

National Research Council. 1995. *Nutrient Requirements of Laboratory Animals: A Report of the Committee on Animal Nutrition*. Washington, DC: National Academies Press.

National Research Council. 2011. *Guide for the Care and Use of Laboratory Animals*. Washington, DC: National Academies Press.

Nelson RJ, Shiber JR. 1990. Photoperiod affects reproductive responsiveness to 6-methoxy-2-benzoxazolinone in house mice. *Biol Reprod* **43**:586.

Nunamaker EA, Otto KJ, Artwohl JE, Fortman JD. 2013. Leaching of heavy metals from water bottle components into the drinking water of rodents. *J Am Assoc Lab Anim Sci* **52**:22.

Pallas BD, Keys DM, Bradley MP, Vernasco-Price EJ, Sanders JD, Allen PS, Freeman ZT. 2020. Compressed paper as an alternative to corn cob bedding in mouse (*Mus musculus*) cages. *J Am Assoc Lab Anim Sci* **59**:496.

Peirson SN, Brown LA, Pothecary CA, Benson LA, Fisk AS. 2018. Light and the laboratory mouse. *J Neurosci Methods* **300**:26.

Perkins SE, Lipman NS. 1996. Evaluation of microenvironmental conditions and noise generation in three individually ventilated rodent caging systems and static isolator cages. *Contemp Top Lab Anim Sci* **35**:61.

Petterborg LJ, Vaughan MK, Johnson LY, ChampneyTH, Reiter RJ. 1984. Modification of testicular and thyroid function by chronic exposure to short photoperiod: a comparison in four rodent species. *Comp Biochem Physiol A* **78**:31.

Rasmussen S, Glickman G, Norinsky R, Quimby F, Tolwani RJ. 2009. Construction noise decreases reproductive efficiency in mice. *J Am Assoc Lab Anim Sci* **48**:263.

Rock ML, Karas AZ, Gartrell Rodriguez KB, Gallo MS, Pritchett-Corning K, Karas RH, Aronovitz M, Gaskill BN. 2014. The time-to-integrate-to-nest test as an indicator of wellbeing in laboratory mice. *J Am Assoc Lab Anim Sci* **53**:24.

Sakellaris PC, Peterson A, Goodwin A, Winget CM, Vernikos-Danellis J. 1975. Response of mice to repeated photoperiod shifts: susceptibility to stress and barbiturates. *Proc Soc Exp Biol Med* **149**:677.

Sales GD, Wilson KJ, Spencer KE, Milligan SR. 1988. Environmental ultrasound in laboratories and animal houses: a possible cause for concern in the welfare and use of laboratory animals. *Lab Anim* **22**:369.

Sanford AN, Clark SE, Talham G, Sidelsky MG, Coffin SE. 2002. Influence of bedding type on mucosal immune responses. *Comp Med* **52**:429.

Särén LE, Hammarberg LK, Kastenmayer RJ, Hallengren LC. 2016: Developing a performance standard for adequate sanitization of wire-bar lids. *J Am Assoc Lab Anim Sci* **55**:765.

Small D, Deitrich R. 2007. Environmental and equipment monitoring. In *The Mouse in Biomedical Research, 2nd ed., Vol. III (Normative Biology, Husbandry, and Models)*, Fox JG, Barthold SW, Davisson MT, Newcomer CE, Quimby FW, Smith AL, Eds. Boston: Academic Press, 409.

Smith BJ, Killoran KE, Xu JJ, Ayers JD, Kendall LV. 2018. Extending the use of disposable caging based on results of microbiologic surface testing. *J Am Assoc Lab Anim Sci* **57**:253.

Steele KH, Hester JM, Stone BJ, Carrico KM, Spear BT, Faith-Goodin A. 2013. Nonsurgical embryo transfer device compared with surgery for embryo transfer in mice. *J Am Assoc Lab Anim Sci* **52**:17.

Suckow MA, Wolter WR, Duffield GE. 2017. The impact of environmental light intensity on experimental tumor growth. *Anticancer Res* **37**:4967.

Taitt KT, Kendall LV. 2019. Physiologic stress of ear punch identification compared with restraint only in mice. *J Am Assoc Lab Anim Sci* **58**:438.

Takeo T, Nakao S, Nakagawa Y, Sztein JM, Nakagata N. 2020. Cryopreservation of mouse resources. *Lab Anim Res* **36**:33.

Taylor JL, Noel P, Mickelsen N. 2019. Evaluation of a 16-week change cycle for ventilated mouse cages. *J Am Assoc Lab Anim Sci* **58**:443.

Tobin G, Stevens KA, Russell RJ. 2007. Nutrition. In *The Mouse in Biomedical Research, 2nd ed., Vol. III (Normative Biology, Husbandry, and Models)*, Fox JG, Barthold SW, Davisson MT, Newcomer CE, Quimby FW, and Smith AL, Eds. Boston: Academic Press, 321.

Toth LA, Trammell RA, Ilsley-Woods M. 2015. Interactions between housing density and ambient temperature in the cage environment: effects on mouse physiology and behavior. *J Am Assoc Lab Anim Sci* **54**:708.

Turner JG. 2020. Noise and vibration in the vivarium: recommendations for developing a measurement plan. *J Am Assoc Lab Anim Sci* **59**:665.

Turner JG, Bauer CA, Rybak LP. 2007. Noise in animal facilities: why it matters. *J Am Assoc Lab Anim Sci* **46**:10.

Vesell ES. 1967. Induction of drug-metabolizing enzymes in liver microsomes of mice and rats by softwood bedding. *Science.* **157**:1057.

Wasson K. 2017. Retrospective analysis of reproductive performance of pair-bred compared with trio-bred mice. *J Am Assoc Lab Anim Sci* **56**:190.

White WW. 2007. Management and design: breeding facilities. In *The Mouse in Biomedical Research, 2nd ed., Vol. III (Normative Biology, Husbandry, and Models)*, Fox JG, Barthold SW, Davisson MT, Newcomer CE, Quimby FW, and Smith AL, Eds. Boston: Academic Press, 235.

Whiteside TE, Thigpen JE, Kissling GE, Grant MG, Forsythe D. 2010. Endotoxin, coliform, and dust levels in various types of rodent bedding. *J Am Assoc Lab Anim Sci* **49**:184.

Willott JF. 2007. Factors affecting hearing in mice, rats, and other laboratory animals. *J Am Assoc Lab Anim Sci* **46**:23.

management

regulatory agencies and compliance

Specific regulatory agencies and requirements may vary with locale. At the time of publication, it is possible for a program in which mice of the genus *Mus* are used for research or teaching to have no regulatory oversight. For example, if the institution receives no funds from the Public Health Service (PHS) for research, the use and care of any mice used would be exempt from the *Public Health Service Policy on Humane Care and Use of Laboratory Animals* (Public Health Service 2015). and from enforcement of the standards of care described in the *Guide for the Care and Use of Laboratory Animals* (the *Guide*) (National Research Council 2011). If the research or testing being conducted will not be used to support the approval process for drugs or medical devices intended for human or animal use, use of the mice will be exempt from the policies described in the *Good Laboratory Practices for Nonclinical Studies* (CFR 21, Food and Drugs). Likewise, if the facility is not accredited by AAALAC International, it is exempt from the enforcement of the standards described in the *Guide*.

Currently, mice of the genus *Mus* bred specifically for use in research are exempt from enforcement of the standards for animal care and use prescribed by the US Department of Agriculture (USDA) as described in the *Regulations of the Animal Welfare Act* (Office of the Federal Register, 9 CFR parts 1,2 and 3 n.d.). In contrast, wild mice such as those of the genus *Mus* not bred specifically for use in research—*Peromyscus* and others—are all regulated under the USDA standards. Readers are cautioned, however, that these regulations are not static and may change in the future to include mice.

DOI: 10.1201/9780429353086-3

Oversight responsibility is described in the Animal Welfare Act (P.L. 91–579, 94–279, 99–198). Registration with USDA and adherence to USDA regulations are required by all institutions, except elementary and secondary schools, using regulated species in teaching, testing, or research in the United States.

Many institutions use mice in activities that do fall under the purview of one or more of the oversight authorities described above. In this regard, the following summary is pertinent.

- **National Institutes of Health, Public Health Service**: Oversight responsibility is described in the Health Research Extension Act of 1985 (P.L. 99–158). The policy is described in the *Public Health Service Policy on Humane Care and Use of Laboratory Animals* (National Research Council 2011). Adherence to the PHS Policy is required of those institutions conducting research using funds from PHS. Principles for implementation of PHS policy are essentially those described in the *Guide for the Care and Use of Laboratory Animals*.

- **US Food and Drug Administration (FDA) and the Environmental Protection Agency (EPA)**: Policies are described in the Good Laboratory Practices for Nonclinical Laboratory Studies (CFR 21 (Food and Drugs), Part 58, Subparts A–K; CFR Title 40 (Protection of Environment), Part 792, Subparts A–L). In general, standard operating procedures must be outlined and rigorously followed and supported with detailed records. Adherence is required when using animals in studies used to request research or marketing permits as part of the approval process for drugs, medical devices, or pesticides.

- **AAALAC International**: AAALAC International is a nonprofit organization designed to provide peer review–based accreditation to animal research programs. The basis for accreditation in the United States is adherence to principles described in the *Guide for the Care and Use of Laboratory Animals*. In the European Union, accreditation is based on adherence to the European Convention for the Protection of Vertebrate Animals Used for Experimental and Other Scientific Purposes (ETS 1,2,3; Council of Europe, 2006). Accreditation is voluntary.

- In addition to the above regulatory bodies, state and local regulations may exist. For example, field studies in which animals are used may come under the purview of a department of natural resources or fish and game authorities.

Institutional Animal Care and Use Committee (IACUC)

The basic unit of an effective animal care and use program is the Institutional Animal Care and Use Committee (IACUC). The USDA, PHS, and AAALAC International require an IACUC at any research institution using animals that falls under their purview.

IACUC Composition

Important points regarding the composition of the IACUC include:

1. Number of members: PHS policy requires a minimum of five members. In contrast, USDA regulations require a minimum of three members.

2. Qualifications of members: The IACUC should include the following:

   - A chairperson.
   - A doctor of veterinary medicine who has training or experience in laboratory animal medicine or science and responsibility for activities involving animals at the research facility.
   - An individual who is in no way affiliated with the institution other than as an IACUC member. At some institutions, this role has been fulfilled by clergy, lawyers, elementary or secondary school teachers, or local humane society or animal shelter officials. In any case, this individual should serve as an advocate for the public perspective.
   - A practicing scientist with experience in animal research (per the PHS Policy).
   - One member whose primary concerns are in a nonscientific area (per the PHS Policy). This individual may be an employee of the institution served by the IACUC.
   - It is acceptable for a single individual to fulfill more than one of the above categories; however, it is generally unacceptable for the attending veterinarian to also serve as the IACUC chair.

Responsibilities of the IACUC

The written regulations should be consulted for an in-depth description of IACUC responsibilities. In general, the IACUC is charged with the following:

- Review proposed protocols for activities involving the use of animals in research, teaching, and testing. Protocols must be approved by the IACUC before animal use may begin.
- Inspect and ensure that the animal research facilities and equipment meet an acceptable standard.
- Ensure that personnel are adequately trained and qualified to conduct research using animals.
- Ensure that animals are properly handled and cared for.
- Ensure that the investigator has considered alternatives to potentially painful or distressful procedures and has determined that the research is nonduplicative.
- Ensure that sedatives, analgesics, and anesthetics are used when appropriate.
- Ensure that proper surgical preparation and technique are utilized.
- Ensure that animals are euthanized appropriately.
- Ensure safety of personnel via oversight of an occupational health and safety program that includes risk assessment and mitigation.

occupational health and zoonotic diseases

Domestic mice purchased from reputable vendors pose virtually no risk of infectious zoonotic disease, unless experimentally infected with zoonotic pathogens. The need for, and aspects of, a comprehensive occupational health and safety program for individuals working with laboratory animals, including mice, have been described (Villano et al. 2017). It is desirable to establish an occupational health and safety program that involves professionals with that specific expertise and that includes focus on risk assessment of personnel and risk mitigation (National Research Council 2011). In general, personnel should cover their street clothing, wear closed-toe shoes, and wear gloves when working with research animals. Occupational health programs for personnel handling mice should be developed with consideration for the following potential health issues:

- **Puncture and bite wounds**: Mice may attempt to bite when they feel threatened or during restraint. Puncture wounds resulting from bites carry the risk of bacterial infection and should be thoroughly cleansed with an antiseptic. In addition, puncture wounds may result from handling equipment with sharp edges or points. For this reason, personnel should be current with respect to tetanus immunization. Needle recapping devices or self-retracting hypodermic needles attached to syringes are commonly employed to minimize the risk of needle-stick injury.

- **Noise**: Facilities in which animals are housed often include areas, such as the cage wash, where significant noise is generated, thus posing risk to personnel of noise-induced hearing loss. In such areas, monitoring of noise levels and development of a hearing conservation program can help mitigate risk to personnel (Randolph et al. 2007). Use of hearing protective devices such as ear plugs or ear muffs can be useful as a means to reduce the exposure of personnel to harmful noise.

- **Allergy**: Allergies to mice are not uncommon in personnel exposed to laboratory animals (Hunskaar and Fosse 1990; Lutsky and Neuman 1975; Schumacher et al. 1981; Yamauchi 1987). Soluble lipocalin proteins (major urinary proteins or MUPs) from the urine and albumin derived from the skin appear to be the major allergens (Krop et al. 2007; Siraganian and Sanberg 1979; Taylor et al. 1977). Further, it has been demonstrated that personnel working with laboratory mice can spread allergens to their home environments (Kube et al. 2020). Personnel may experience respiratory symptoms such as sneezing and rhinitis or skin symptoms such as redness, swelling, and pruritis following exposure. As with many allergies, extreme allergic sensitivity to mice can result in anaphylaxis and thus represents a serious occupational hazard for some individuals. If animals are handled outside of animal transfer stations or biosafety cabinets, respiratory protection for personnel may be warranted, but gloves and a disposable gown are always advised. If fitted respirator use is required, such devices should be fit-tested for each individual. One study found that dust respirators may vary from 65% to 98% in efficiency related to the exclusion of mouse allergens (Sakaguchi et al. 1989; Slovak et al. 1985). Alternatively, ventilation and building designs, such as down-ventilated benches and ventilated cages, can be used to decrease the exposure of personnel to animal allergens. The use of filter tops on cages, caging systems that ventilate to building exhaust ducts, ventilated cage changing stations, ventilated bedding disposal systems, and corncob bedding also appears to reduce the level of mouse allergens (Feistenauer et al. 2014; Langham et al. 2006). Ideally, sensitive personnel should be reassigned to job tasks that minimize the possibility of exposure to allergens. Personnel with the greatest risk of allergen exposure are those who perform cage changing and emptying cages of bedding (Feistenauer et al. 2014). The advice of an occupational health specialist should be sought and followed if reassignment away from mouse areas is not possible. In addition, it is advisable for such individuals to undergo periodic respiratory function testing.

- **Experimental biohazards**: Some studies may involve purposeful infection of mice with known human pathogens. In such cases, it is recommended that standard operating procedures for safe handling of biohazardous materials and infected animals be established and followed. Guidelines for the use of biohazardous agents are presented in detail elsewhere (Centers for Disease Control and Prevention, National Institutes of Health (CDC/NIH) 2020). The rise of humanized mouse models and the implantation of neoplastic tissue of human origin raises the possibility of permissive infections of mice with human viruses. Human tissues for transplantation into mice should be screened for both murine and human infectious agents. Likewise, appropriate consideration should be given to studies in which chemical or radiologic exposure of personnel may occur.

- **Hantavirus**: The genus *Hantavirus* is a member of the family Bunyaviridae. Infection with Old World hantavirus is associated with severe hemorrhagic disease with renal involvement in humans, while New World hantaviruses are associated with respiratory distress syndromes. Hantavirus does not appear to cause any illness in rodents (Kawamura et al. 1991). The virus is shed in the saliva, urine, and feces of infected mice. Most commonly, hantavirus is isolated from wild rodents such as those encountered during field studies or, possibly, verminous rodents (Hankenson et al. 2003). Mice from reputable commercial sources should be free from hantavirus; however, the possibility of infection exists, albeit small, and is one reason why those working with mice outside of biocontainment systems should at least wear dust masks and gloves when handling mice or their waste products.

- **Lymphocytic choriomeningitis**: An *Arenavirus*, lymphocytic choriomeningitis virus (LCMV) generally produces an asymptomatic infection in mice (Hotchin 1971). It should be noted that hamsters may also have LCMV as an unapparent infection (Parker et al. 1976). Commercial colonies are generally free of LCMV; however, experimental transplantation of tumors into mice may serve as a route for the passage of LCMV. LCMV may be transmitted to humans by direct contact with feces and urine, inhalation of dried excreta, or by the bite of an infected mouse. In humans, the severity of the disease is variable but is usually described as having symptoms typical of a mild case of influenza. In severe cases, the central nervous system may be affected. As with many other diseases, LCMV may cause damage to the fetus if a pregnant person is infected.

- **Dwarf tapeworm**: The cestode *Rodentolepis* (formerly *Hymenolepis*) *nana* infrequently infects laboratory-bred research mice. Most infestations in mice are subclinical; however, the parasite is zoonotic and may result in disease in humans.

compassion fatigue

Personnel working with animals sometimes experience compassion fatigue, an emotional exhaustion that can occur following repeated and prolonged work that involves a sense of being overwhelmed by the conflict between having to perform tasks, such as euthanasia, and the compassion for animals that personnel might feel as part of the human–animal bond. Those affected might experience symptoms that include sadness and apathy, physical and emotional tiredness, and others. Those managing programs that require euthanasia or other procedures that might result in pain or distress to animals should be aware of this possibility and consider offering support to personnel who experience compassion fatigue (Newsome 2019).

references

Centers for Disease Control and Prevention, National Institutes of Health (CDC/NIH). 2020. *Biosafety in Microbiological and Biomedical Laboratories*, 6th ed. Atlanta: CDC/NIH.

Feistenauer S, Sander I, Schmidt J, Zahrandik E, Raulf M, Brielmeier M. 2014. Influence of 5 different caging types and the use of cage-changing stations on mouse allergen exposure. *J Am Assoc Lab Anim Sci* **53**:356.

Hankenson FC, Johnston NA, Weigler BJ, Di Giacomo RF. 2003. Zoonoses of occupational health importance in contemporary laboratory animal research. *Comp Med* **53**:579.

Hotchin J. 1971. The contamination of laboratory animals with lymphocytic choriomeningitis virus. *Am J Pathol* **64**:747.

Hunskaar S, Fosse RT. 1990. Allergy to laboratory mice and rats: a review of the pathophysiology, epidemiology and clinical aspects. *Lab Anim* **24**:358.

Kawamura K, Zhang XK, Arikawa J. 1991. Susceptibility of laboratory and wild rodents to *Rattus* or *Apodemus*-type hantaviruses. *Acta Virol* **35**:54.

Krop EJM, Matsui EC, Sharrow SD, Stone MJ, Gerber P, van der Zee JS, Chapman MD, Aalberse RC. 2007. Recombinant major urinary proteins of the mouse in specific IgE and IgG testing. *Int Arch Allergy Immunol* **144**:296.

Kube H, Herrera R, Dietrich-Gümperlein G, Schierl R, Nowak D, Radon K, Wengenroth L, Gerlich J. 2020. From workplace to home environment: spreading of mouse allergens by laboratory animal workers. *Int Arch Occup Environ Health*, https://doi.org/10.1007/s00420-020-01603-9.

Langham GL, Hoyt RF, Johnson TE. 2006. Particulate matter in animal rooms housing mice in microisolation caging. *J Am Assoc Lab Anim Sci* **45**:44.

Lutsky I, Neuman I. 1975. Laboratory animal dander allergy. I, An occupational disease. *Ann Allergy* **35**:201.

National Research Council. 2011. *Guide for the Care and Use of Laboratory Animals.* Washington, DC: National Academy Press.

Newsome JT, Clemmons EA, Fitzhugh DC, Gluckman TL, Creamer-Hente MA, Tambrallo LJ, Wilder-Kofie T. 2019. Compassion fatigue, euthanasia stress, and their management in laboratory animal research. *J Am Assoc Lab Anim Sci* **58**:289.

Office of the Federal Register, 9 CFR parts 1,2 and 3. n.d. *Fed Regist* **54**(168):36133.

Parker JC, Igel HJ, Reynolds RK, Lewis AM, Jr, Rowe WP. 1976. Lymphocytic choriomeningitis virus infection in fetal, newborn, and young adult Syrian hamsters. *Infect Immun* **13**:967.

Public Health Service. 2015. *Public Health Service Policy on Humane Care and Use of Laboratory Animals.* Washington, DC: US Department of Health and Human Services.

Randolph MM, Hill WA, Randolph BW. 2007. Noise monitoring and establishment of a comprehensive hearing conservation program. *J Am Assoc Lab Anim Sci* **46**:42.

Sakaguchi M, Inouye S, Miyazawa H, Kamimura H, Kimura M, Yamazaki S. 1989. Evaluation of dust respirators for elimination of mouse aeroallergens. *Lab Anim Sci* **39**:63.

Schumacher MJ, Tait BD, Holmes MC. 1981. Allergy to murine antigens in a biological research institute. *J Allergy Clin Immunol* **68**:310.

Siraganian RP, Sanberg AL. 1979. Characterization of mouse allergens. *J Allergy Clin Immunol* **63**:435.

Slovak AJM, Orr RG, Teasdale EL. 1985. Efficacy of the helmet respirator in occupational asthma due to laboratory animal allergy (LAA). *Am Ind Hyg Assoc J* **46**:411.

Taylor AN, Longbottom JL, Pepys J. 1977. Respiratory allergy to urine proteins of rats and mice. *Lancet* **2**:847.

Villano JS, Follo JM, Chappell MG, Collins Jr, MT. 2017. Personal protective equipment in animal research. *Comp Med* **67**:203.

Yamauchi C. 1987. The survey of laboratory animal allergy in Japan: study group on the present state of laboratory animal allergy and its countermeasures. *Jikken Dobutsu* **36**:95.

clinical medicine

basic veterinary supplies

The following basic supplies are useful for the clinical care and experimental manipulation of mice.

1. Disposable syringes, 1 ml and 3 ml.
2. Disposable hypodermic needles, particularly 25–26 gauge (diameter) and $\frac{5}{8}$ inch (length).
3. Gauze sponges and cotton swabs.
4. Disinfectant, such as povidone-iodine or dilute chlorhexidine.
5. Topical antibiotics, such as triple antibiotic ointment or ophthalmic gentamicin ointment, with and without dexamethasone, including ophthalmic ointments.
6. Sterile fluids for SQ administration, such as lactated Ringer's solution or 0.9% sodium chloride. Phosphate-buffered saline (PBS) is not suitable for parenteral administration.
7. 50% dextrose for oral administration.
8. Bacterial culture swabs in transport media.
9. Several 22-gauge ball-tipped, stainless steel or disposable feeding needles for orogastric gavage.
10. Fluorescein stain and cobalt blue light.
11. Scissors.
12. Ear tag cutters.

DOI: 10.1201/9780429353086-4

13. Additional supplies should supplement those listed above, depending upon the needs of the facility.

---

physical examination of the mouse

Physical examination should be performed on mice upon arrival at the facility and on mice exhibiting any abnormalities. With large groups of mice, a few representative individuals or any obviously abnormal animals should be examined. Findings should be recorded in an appropriate health record. Physical examination of the mouse can be performed in the following manner.

1. General assessment of behavior of the animal within the cage and during removal from the cage should be performed. Mice are inquisitive and will generally be observed moving about the cage, particularly after disturbance such as cage movement. Findings such as lethargy, aggressiveness, poor nest building, isolation from cagemates, abnormal movements, or hunched appearance should be noted.

2. The coat should be examined for hair loss, open or closed skin lesions, and abnormal masses. In addition, the overall quality of the hair coat should be assessed, since an unkempt appearance is often evidence of underlying illness. Skin wounds, including those that are healing, may indicate fighting among the mice, which might require separation of individuals to prevent further injury.

3. Overall body condition should be evaluated and can be a more reliable indicator of disease than weight, as some health conditions can cause increases in body weight while breaking down muscle and fat (Burkholder et al. 2012). Body condition can be assessed by passing a finger over the spine and hip bones and assigning a score from 1 to 5, with 1 being emaciated and 5 being obese. Optimal body condition is a 3 (Ullman-Cullere and Foltz 1999). Mice that are thin or abnormally small compared with littermates may have underlying illness.

4. Attention should be paid to any "chattering" noises as evidence of possible respiratory disease. Chattering is a stertorous breathing noise presumably due to mucopurulent material in the airways and/or nose.

5. The eyes, nose, ears, and perineal region should be examined for discharges, swellings, or other abnormalities.

6. The incisors should be examined for overgrowth.

7. Palpation for abnormal masses within the abdominal cavity can be performed by restraining the mouse by the scruff of the neck (Figure 4.1) and firmly pressing the thumb and index finger of the free hand into the cranial part of the abdomen and slowly drawing the fingers back caudally, being sure to palpate both ventral and caudal aspects of the abdomen.

8. Body temperature may be measured via several methods in mice. Infrared temperature measurement is least invasive but may not provide consistent accuracy in furred mice (Fiebig et al. 2018). Rectal temperature in mice should be taken using probes designed specifically for the species, as the probe size of standard thermometers may cause soft-tissue damage. Surgically implanted telemetry devices allow for accurate, real-time temperature monitoring, although the invasiveness of this method typically limits it to research use. The normal body temperature for mice may vary slightly depending

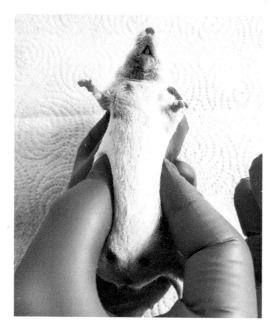

Figure 4.1 Mouse palpation. Abdominal palpation of the mouse (ventral side shown).

upon the strain and the method of measurement; however, temperatures in the range of 36.5–38°C (97.5–100.4°F) are typical (Kort et al. 1998). Body temperature can be useful for predictions of survival in severely ill mice, as prolonged hypothermia has been shown to be predictive of death in multiple survival studies (Cates et al. 2014; Toth 2018).

common spontaneous and noninfectious diseases

Different spontaneous disease conditions are more likely to manifest at different ages (Table 4.1) and can also vary with sex, genetic background, immune status, microbial status, diet, and other factors. Sexual dimorphisms in body weight, body size, and the morphology of salivary glands, adrenals, kidneys, and mammary glands should be recognized as such and not interpreted as important findings. However, the absence of sexual dimorphisms may be a significant finding.

Malocclusion and hydrocephalus are life-threatening conditions that should be identified at or before weaning. Congenital eye conditions may be identified early in life; they are not life-threatening but may interfere with some studies. The most significant (life-threatening or likely to interfere with research) conditions in adult animals are dermatitis, wounds from fighting, abdominal enlargement for various reasons other than pregnancy, and neurologic signs. Any mice that suffer a decline in body condition should be evaluated for illness. Likely noninfectious causes of progressive decline, or wasting, in older mice (more than 6 months old in some strains, more than 12 or 18 months

TABLE 4.1: NONINFECTIOUS CLINICAL CONDITIONS IN MICE AT DIFFERENT AGES

	Young Mice	Adult Mice	Older Mice
Life-threatening	Hydrocephalus Malocclusion	Dermatitis and wounds Abdominal enlargement (e.g., hydrometra, hydronephrosis, urinary obstruction, ascites) Neurologic signs (e.g., paresis-paralysis, seizures, vestibular signs)	Neoplasia (tumors) Wasting due to chronic progressive conditions (e.g., amyloidosis, nephropathy, neoplasia)
Usually not life-threatening	Microphthalmia Anophthalmia Congenital cataract	Alopecia due to barbering Hearing loss Corneal opacity	Obesity

old in long-lived strains) include systemic amyloidosis, severe renal disease, acidophilic macrophage pneumonia, and neoplasia. Arteritis (polyarteritis), mild cardiac changes, ectopic mineralization, and hyalinosis are also likely in old mice, but are usually not life-threatening. Obesity in overfed older animals is a management problem, but also may reflect underlying genetic predisposition. Neoplasms should be expected in aging mice, and mouse strains vary in the tumor types that are likeliest to develop. Knowledge of genetic background should inform what tumors and other phenotypes to expect, and what phenotypes may be unusual or important. The most common neoplasms reported in common mouse strains involve the hematopoietic system (lymphomas and histiocytic sarcoma), lungs, mammary glands, and liver. But, like other species, mice may develop neoplasms in almost any tissue. Some expected neoplasms and other expected or likely phenotypes in common strains are summarized in Table 4.2.

TABLE 4.2: NONINFECTIOUS CONDITIONS AND NEOPLASMS, BY MOUSE STRAIN

Mouse Strain	Condition
129 Strains	Acidophilic macrophage pneumonia, hyalinosis Lung tumors
A/J	Congenital anomalies; amyloidosis; muscular dystrophy Lung tumors
AKR	Thymic lymphoma
BALB/c	Conspecific (male) aggression; cardiac calcinosis; cardiac thrombi Lung tumors; Harderian gland tumors; myoepithelioma; induced plasmacytoma
C3H	Blind ($Pde6b^{rd1}$); cardiac calcinosis and other soft-tissue mineralization; mammary tumors in females; liver tumors in males
C57BL/6	Hydrocephalus; micro- and anophthalmia; ulcerative dermatitis; amyloidosis; acidophilic macrophage pneumonia; hyalinosis Lymphoma; histiocytic sarcoma
DBA/2	Cardiac calcinosis and other soft-tissue mineralization; seizures; deafness; glaucoma
FVB/N	Blind by weaning ($Pde6b^{rd1}$); seizures; mammary hyperplasia Lung tumors
NOD	Diabetes; immune alterations Lymphoma
SJL/J	Blind by weaning ($Pde6b^{rd1}$); conspecific (male) aggression; muscular dystrophy Lymphoma
Swiss outbred	Stock source variations in: blind by weaning ($Pde6b^{rd1}$); acidophilic macrophage pneumonia; amyloidosis; nephropathy; urinary syndrome Lymphomas; lung tumors; liver tumors; skin tumors

Common Conditions Involving the Alimentary System

Non-neoplastic conditions

Periodontal inflammation involving molar teeth, sometimes with hair from the pelt visible, is usually an incidental finding that does not contribute to morbidity or mortality. Especially in older mice, the inflammation can be substantial and accompanied by alveolar bone loss and remodeling.

Incisor dysplasia (abnormal or disrupted incisor growth) is usually an incidental histopathology finding not associated with clinical signs or disease. The condition is more common in older mice and has been associated with feeding of soft or powdered food. Primary tumors of the teeth such as odontomas can occur but are uncommon in most mouse strains.

At (or before) weaning, mice should be examined for overgrowth and misalignment of incisor teeth that will prevent these mice from eating hard food (Figure 4.2). Mice that fail to thrive after weaning should also be examined for the condition. Usually, these mice should be culled from a breeding or research program. If they are genetically valuable, they may be maintained by regular trimming, but the teeth may be damaged by the procedure and develop chronic infections.

Figure 4.2 Malocclusion of the incisors in a C57BL/6 mouse.

Since mice have continuously growing teeth, they can also develop malocclusion later in life if the incisors are damaged. This may happen if animals bite wire lids as they are being restrained or may be associated with tooth placement in stereotaxic apparatus.

Esophageal dilatation, or megaesophagus, is sometimes identified at necropsy. When severe, it may be a cause of death.

Alimentary tumors

Spontaneous, primary tumors of the intestine and pancreas are not common in most mouse strains. Nor are spontaneous, primary tumors of the salivary glands, although primary adenomas, carcinomas, or myoepitheliomas of the salivary gland are possible. Papillomas in the stomach are sometimes reported, usually as incidental findings.

Spontaneous, primary neoplasms of hepatocytes are expected with variably high incidence in aging male mice of certain strains, especially C3H, CBA, and B6C3F1. There may be single or multiple nodules. Hepatocellular adenoma and hepatocellular carcinoma are more likely than hepatoblastoma. Cholangioma, cholangiocarcinoma, Ito cell tumors, hemangiomas, hemangiosarcoma, and metastatic neoplasms may also be seen in mouse livers.

Common Conditions Involving the Cardiovascular System

Non-neoplastic conditions

Arteritis, polyarteritis, and periarteritis are usually incidental histopathology findings not associated with clinical signs or disease. These conditions are more likely in older mice. Arteries at multiple anatomic sites can be involved (polyarteritis). Periarterial inflammation and fibroplasia may be substantial in advanced disease (periarteritis). Arteries in the mesentery, pancreas, heart, head, and other sites may be affected. Severe involvement of the heart and around the brain may contribute to morbidity or mortality.

Cardiac thrombi (intravascular blood clots that form in vivo) usually involve the left atrium in mice. They are not especially common in most unmanipulated mice, but are more common in BALB/c mice, along with cardiac calcinosis and degenerative myocardial changes. Small thrombi may be identified by histology without apparent clinical significance. When large thrombi are accompanied by cardiomegaly, cardiac dilatation, or hypertrophy, it may be difficult to determine which condition is primary.

Mineralization of cardiac myofibers or epicardium is expected in the related BALB/c, C3H, and DBA/2 mouse strains, but is unusual in most other strains. In BALB/c, mineralization commonly involves the epicardial surface of the right ventricular free wall (Figure 4.3). A candidate gene has not been identified in BALB/c mice, and the incidence differs between substrains (Glass et al. 2013). In C3H and DBA/2 mice, other areas of the myocardium and other soft tissues, including tongue, may be affected. The gene involved in this dystrophic mineralization is the *Abcc6* gene (Aherrahrou et al. 2008).

Histopathology findings, including myofiber hypertrophy, degeneration, loss, and/or replacement fibrosis, sometimes accompanied by inflammation, are fairly common in some strains of mice and are likely to increase with age. These and other cardiac changes, including cardiac thrombi and calcinosis, sometimes have been referred to collectively as cardiomyopathy.

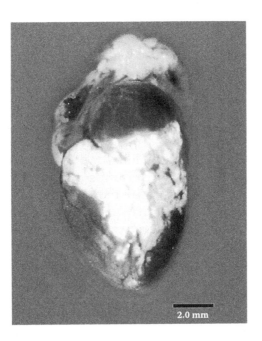

2.0 mm

Figure 4.3 Epicardial mineralization in an 11-month-old BALB/c mouse. The white (mineral) material covers the epicardial surface and can infiltrate the underlying myocardium. In this strain, the right ventricular free wall typically is affected earliest and most severely. DBA/2 and C3H strains are also susceptible to this condition, which may involve deeper myocardium and other soft tissues.

Cardiovascular tumors

Hemangioma and hemangiosarcoma are benign and malignant, respectively, and are primary neoplasms of the vascular system. They are not especially common in most mouse strains but are encountered in various tissues in aging mice on long-term studies. Especially in highly vascular tissues such as liver and spleen, it can be challenging to distinguish these neoplasms from large areas of angiectasis (sometimes referred to as telangiectasis or peliosis). Angiectasis refers to abnormally dilated vascular spaces and usually is an incidental histopathology finding in older mice. Bloody tumor masses in liver or spleen, or less commonly in skin or other tissues, suggest hemangioma or hemangiosarcoma.

Common Conditions Involving the Endocrine System

Non-neoplastic conditions

Small non-neoplastic accessory adrenal cortical nodules are common in mice of some strains and various ages. In female mice, vacuolar degeneration of the perimedullary X zone can be a conspicuous histology finding.

The thyroid and parathyroid glands and the thymus all develop from embryonic structures called *pharyngeal pouches* in the embryonic pharynx. Fragments of thyroid or parathyroid may fail to migrate completely and can be identified by histopathology on the midline between the intrathoracic thymus and their usual location near the larynx. Fragments of thymus may be found on the midline beyond the thorax, or near the thyroid and parathyroid glands. Ectopic fragments of these structures usually have no clinical significance. However, residual activity from the ectopic fragments can frustrate studies in which these tissues were thought to have been ablated or removed.

Endocrine tumors

Adrenal cortical subcapsular cell hyperplasia is a common finding in aged mice, and adrenal cortical tumors may be found frequently in some strains, usually as incidental findings (Shinohara and Frith 1986). Adrenal medullary tumors (pheochromocytomas) are less common in mice. Pancreatic islet tumors (insulinomas, islet cell tumors) are unusual in mice, but large or hyperplastic islets may be identified in obese mice and in some mouse models of diabetes. Pituitary neoplasms can be common in long-term studies of aging mice. These tumors are usually identified by histologic examination and are not obvious on gross examination unless lesions are large and the head is examined carefully. Proliferative and neoplastic pituitary lesions in mice involve the pars distalis most commonly, may secrete prolactin, and may be associated with proliferative mammary lesions in some strains, especially FVB/N (Radaelli et al. 2009). Spontaneous thyroid tumors are uncommon. When they do occur, follicular cysts, or follicular hyperplasia and tumors, are more common than interstitial cell (C cell) proliferative lesions. Hypothyroidism and proliferative changes can be induced by treatment with high doses of the antibiotic trimethoprim-sulfamethoxazole (Altholtz et al. 2006).

Hematopoietic and Immune System

Non-neoplastic conditions

Anemia refers to a reduced number of circulating red blood cells. Iatrogenic anemia related to blood sampling for research purposes may be the most common cause of this condition in research mice. Anemia due to reduced production of red blood cells can occur when hematopoietic tissue is responding to a severe or chronic infection, when hematopoietic neoplasms take over much of the bone marrow, or there is damage to marrow by toxins or irradiation. Pallor (pale paws, ears, and eyes), "watery" blood, and failure of blood to clot suggest severe compromise to hematopoietic tissue. Anemia due to primary destruction of red blood cells (hemolytic disease) or due to a primary failure of red blood cell production is possible but is not likely in common inbred strains.

Reactive hyperplasia of inflammatory cell precursors followed by increased numbers of circulating leukocytes are expected responses to infection in mice. Increased number of circulating neutrophils is expected in bacterial infections. Increased number of circulating eosinophils is expected with parasitism. Increased numbers of lymphocytes and/or monocytes are expected in chronic infections. Characteristic gross findings include enlarged spleen (splenomegaly) and lymph nodes (lymphadenomegaly) near affected sites. Characteristic histopathology findings include increased immature and mature granulocytes in bone marrow, spleen, and sometimes liver; and increased immature and mature lymphocytes and plasma cells in enlarged lymph nodes and spleens. Extreme reactive proliferative responses sometimes can be difficult to distinguish from hematopoietic neoplasia. Identification of the infection can help characterize the proliferation as reactive (inflammatory) or neoplastic.

The hematopoietic and immunopoietic systems respond to stress, and to inanition or starvation. Stress responses in mice (and rats) are mediated largely by corticosterone, which induces lymphocyte death. Lymphocyte depletion or loss in lymphoid organs (thymus, spleen, lymph nodes) may be a manifestation of stress from various sources including disease, transportation, starvation, and other environmental factors.

Hematopoietic and immune system tumors

Hematopoietic neoplasms are common in mice and are the likeliest causes of death in several strains. Thymic lymphoma is a likely cause of death in AKR, C58, NOD, SCID, nude, and related strains before 1 year of age. Other lymphoma types and histiocytic sarcoma usually occur in older mice and are a significant cause or contributor to death in SJL/J and C57BL/6 among others. In advanced disease, some lymphomas and histiocytic sarcomas can involve bone marrow and circulation in a leukemic phase, but primary leukemias, myeloid neoplasms, and mast cell tumors are uncommon in standard inbred mouse strains. Marked leukocytosis (high white cell count), especially neutrophilic leukocytosis, in mice is much more likely to be due to infection than to leukemia. Advanced or disseminated hematopoietic neoplasia is likely to present clinically as a decline in body condition. Peripheral lymphadenomegaly (Figure 4.4) may be palpable or obvious as enlarged symmetric neck, axillary, or inguinal masses. Enlarged thymus, lymph nodes, spleen, or liver usually are obvious at necropsy (Figure 4.5). With advanced disease and severe organomegaly, a "normal" body weight may be misleading due to neoplastic infiltration and enlargement of lymph nodes, spleen, and other organs. Pallor (pale paws, ears, and eyes) and watery blood are good indicators that the marrow is affected and unable to produce red cells.

Common Conditions Involving the Integumentary System

Non-neoplastic conditions

Barbering by one or more mice on their cagemates is a likely cause of symmetric patterns of alopecia (baldness). Hair removal is usually

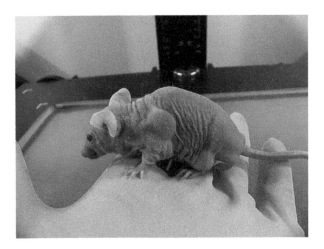

Figure 4.4 Nude mouse with peripheral lymphadenopathy, a finding often associated with lymphoma.

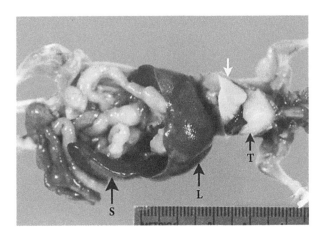

Figure 4.5 Thymic lymphoma in a mouse. The thymus (T) is massively enlarged, almost as large as the lungs (white arrow). There are hepatomegaly (L) and splenomegaly (S), also due to lymphoma.

by nibbling, without associated skin trauma. Vibrissae (commonly, but incorrectly, called whiskers) are important sensory organs in mice, using as much cortical processing power in the mouse brain as human hands do in the human brain. When whiskers are removed by barbering, mice may perform differently in behavioral tests.

Ulcerative dermatitis (Figure 4.6) is usually progressive and recurring and can interfere with many types of research. Treatment

Figure 4.6 Ulcerative dermatitis in a C57BL/6 mouse. Although it is more common in black mice of C57BL/6 origin, ulcerative dermatitis can be a problem in other strains of mice. Frequently, the back of the head and neck are affected. This mouse was negative for mites.

modalities have included various topical salves and dietary interventions, but the simplest treatment that seems to give good results is trimming the toenails (Adams et al. 2016). Severe progressive ulcerated lesions also can be life-threatening (from overwhelming opportunistic infections) or can necessitate euthanasia for humane reasons.

Skin wounds due to fighting usually are seen on the back, rump, and around the genitals (Figures 4.7 and 4.8). There is usually one unwounded animal that is the source of the trouble. Less commonly, animals may show wounding patterns to the axillary regions. With this pattern, all animals in the cage are typically wounded. The full extent of the wounds may not be obvious from casual observation, but necrotic dry skin may overlie extensive wounds. Aggression, particularly among males, can be a problem in many strains, including BALB/c and SJL/J.

Tumors of skin and adnexae

While primary spontaneous skin neoplasms are not very common in most mouse strains, papillomas, carcinomas, and subcutis sarcomas can occur. In susceptible strains, these can be induced by ear tags or implanted materials such as transponders.

Figure 4.7 Fight wound on the back just above the tail base of a male mouse. Wounds may be open and fresh wounds or crusty, healing wounds such as in this mouse.

Figure 4.8 Preputial fight wounds in C57BL/6 male mice. The wounds are the open lesions around the genitalia.

Clitoral or preputial gland neoplasms are possible causes of enlarged clitoral or preputial glands, but tumors are far less likely than adenitis of these structures. In older mice, these glands may become cystic, with mild proliferative changes, usually associated with inflammation.

Harderian gland neoplasms generally are uncommon, but are more likely in BALB/c mice (Sheldon et al. 1983). Large tumors may present clinically as exophthalmos (protruding eye) due to a mass behind the eye. Orbital or retro-orbital abscess is an important diagnostic consideration for exophthalmos.

Mammary adenoma and carcinoma are the most common spontaneous tumors of adnexae, especially in aging female C3H and GR mice. Exogenous mammary tumor virus (Bittner agent) contributed to incidences of close to 100% in the past (e.g., before 1980), but this agent has been eliminated from most contemporary colonies. Mouse mammary tissue is widely distributed from the tail to the ears in females, and so neoplasms may arise almost anywhere in the subcutis. FVB/N female mice develop mammary hyperplasia that may complicate mammary tumor studies using this strain (Radaelli et al. 2009). Myoepitheliomas may arise in any tissue with myoepithelial cells, but especially in salivary and mammary glands, and are more common in BALB/c than in other strains.

Common Conditions Involving the Musculoskeletal System

Non-neoplastic conditions

Arthritis (inflammation of, or in, joints) occurs spontaneously in some mouse strains and can increase with age and obesity. Infectious causes of arthritis should be suspected in apparent outbreaks of this condition, especially in strains where spontaneous arthritis is not expected (Bremell et al. 1990). *Fibro-osseous lesion* describes a variably common lesion in the medullary cavities of aging, usually female, mice. In various bones, hematopoietic tissue can be mildly to markedly replaced by fibrovascular proliferation with or without osseous (bony) contributions. Usually, this is an incidental finding.

Muscular dystrophy due to mutations in the dysferlin gene occurs in A/J (*Dysf^{prmd}*) and SJL/J (*Dysf^{im}*) mice. Clinical signs of weakness may be subtle until the mice are challenged in neurobehavioral tests. Histopathology findings of myofiber degeneration, regeneration, and fibrosis increase with age. Other spontaneous and genetically induced models are used to model human muscular dystrophies. Heart muscle can be involved in some of these models.

Bone quality, density, mass, and strength vary among mouse strains and can be influenced by factors such as diet, obesity, and estrogen levels. Spontaneous fractures in mice are unusual and suggest a mutation affecting bone.

Musculoskeletal tumors

Spontaneous, primary neoplasms of smooth muscle—leiomyoma (benign) and leiomyosarcoma (malignant)—or of skeletal muscle—rhabdomyoma (benign) and rhabdomyosarcoma (malignant)—are uncommon in mice. Smooth muscle neoplasms usually are reported in the uterus or urinary bladder, where they can be difficult to distinguish from other mesenchymal (nonepithelial) tumors. Rhabdomyosarcomas presenting as subcutaneous masses of the trunk or limbs have been reported in A/J and BALB/c mice.

Spontaneous, primary neoplasms of bone—osteoma (benign) and osteosarcoma (malignant)—are also uncommon in mice. Occasional occurrence of multiple osteomas, especially in the skull and larger limb bones, has been associated with retroviruses. Osteosarcomas can be common in some *Trp53*-deficient and BALB/c mice.

Common Conditions Involving the Nervous System

Non-neoplastic conditions

Absent or reduced corpus callosum (normal neural connection between brain hemispheres) is expected (with variable incidence) in mice of BALB/c and 129 strains. Associated disease signs are uncommon. The condition should be recognized as a possibly normal anatomic variation for the genetic background when it is identified by dissection, imaging, or histology.

Dark neurons and vacuolation of the white matter occur as artifacts related to postmortem handling of tissues. Dark neurons can result from handling of unfixed central nervous system (CNS) tissue and should not be confused with dark necrotic neurons. Vacuolation artifact usually occurs in white matter, sometimes with pale basophilic acellular material in the vacuoles. It is frequently associated with prolonged exposure to alcohols during fixation and processing. These vacuoles should not be confused with spongiosis or edema of the neuropil, or with distended myelin sheaths as in demyelinating conditions. For CNS studies, it can be very important for specimens from experimental and control mice to be dissected, handled, and processed identically or concurrently.

Many common laboratory mouse strains, including C3H, FVB/N, SJL/J, SWR, and some outbred Swiss mice, are blind due to homozygosity for the recessive allele *rd1* (in the *Pde6b* gene), which causes retinal degeneration and loss of photoreceptors by the time of weaning. Mice with microphthalmia have variably severe abnormalities in retina, lens, and other structures. DBA/2 mice develop glaucoma as they age.

Cataracts and corneal opacities can present similarly as pale areas within or on the anterior globe. Careful examination should reveal whether the abnormality is in the lens (in the eye) or the cornea (the anterior surface). Opacities of the lens (cataracts) may be genetically determined or may be induced by irradiation or other treatments, and some may be transient. Corneal opacities can occur in various strains, but are more likely in C3H, DBA, BALB/c, and related strains.

At (or before) weaning, mice should be examined for a domed and relatively enlarged head that indicates dilated brain ventricles (*hydrocephalus*) (Figure 4.9). These mice usually are smaller than unaffected littermates and are likely to do poorly and die. Affected mice should be culled from breeding or research populations as they are unlikely to be good candidates for either, as well as there being associated welfare issues. The condition is most commonly seen in C57BL/6 or C7BL/10 strain mice. With later onset hydrocephalus, the head is not enlarged or domed, because cranial sutures closed before the onset of ventricle enlargement. Markedly enlarged ventricles can be identified postmortem when the head or brain (fresh or

Figure 4.9 C57BL/6 mouse with domed head typical of hydrocephalus.

fixed) is sectioned. Ventricle size varies among strains, with C57BL/6 having relatively large ventricles. Mild enlargement identified with imaging techniques may not be as obvious in postmortem specimens if the cortex collapses in after the cerebrospinal fluid (CSF) drains from the ventricles.

At (or before) weaning, micro- or anophthalmia can be recognized as small or inapparent eyes (Figure 4.10). The condition is reported to be most common in C57BL/6 and related mice, in females, and in the right eye. It is generally desirable to cull these mice from a breeding program.

Abnormal gait, posterior weakness (*paresis*), or inability to move limbs may suggest a primary neurologic problem, or possibly a musculoskeletal problem, trauma, or other illness.

Clinical signs such as rolling, spinning, or head tilt usually indicate damage to the vestibular system. Otitis interna, arteritis, infarcts, or other lesions of, or near, the vestibular nuclei or nerve tracts are possible causes. Infectious causes of otitis and upper respiratory tract infections are discussed later in this chapter.

Nervous system tumors

Spontaneous, primary neoplasms of the nervous system in mice are uncommon. Neoplasms involving the spine or head, such as osteosarcoma or hematopoietic neoplasms, may present with neurologic signs due to compression or invasion of spinal cord or brain.

Figure 4.10 C57BL/6 mice with normal appearance (left) and anophthalmia (right).

Common Conditions Involving the Respiratory System

Non-neoplastic conditions

Acidophilic macrophage pneumonia or *eosinophilic crystalline pneumonia* can be a common and important cause of death in susceptible mouse strains, such as 129, C57BL/6, Swiss, and related mice. In advanced disease, lungs can be grossly consolidated and pale, with corroborating histopathology findings of engorged acidophilic macrophages and crystals filling airways. In mild conditions, scattered macrophages laden with eosinophilic granular or crystalline material and scattered extracellular eosinophilic crystals may be an incidental histopathology finding. Infectious pneumonia conditions can accompany acidophilic macrophage pneumonia. The acidophilic material has been identified as a chitinase-like protein (YM1) that is associated with various immune-stimulated conditions. When these chitinase-like proteins are produced by epithelia at other anatomic sites, usually in older mice, the condition has been referred to as *hyalinosis* (see section titled "Common Systemic or Multisystem Conditions").

Respiratory tumors

Spontaneous, primary lung tumors are common and a likely cause or contributor to death in certain strains, especially A, 129, BALB/c, and FVB. Multiple tumors, usually more adenomas than carcinomas, can occur before 1 year of age in susceptible strains. Metastatic involvement of lungs by liver, mammary, hematopoietic, or other neoplasms may occur. Dyspnea and clinical deterioration may be evident in severely affected mice.

Common Conditions Involving the Urogenital System

Non-neoplastic conditions

Female mice have a vaginal closure membrane at birth that should dissolve at puberty. If the membrane fails to dissolve, imperforate vagina results in progressive enlargement of the uterus with fluid or mucoid material (*hydrometra* or *mucometra*). Progressive abdominal swelling, often with a bulging perineum, and infertility are characteristic clinical findings. *Pyometra* (pus in the uterus) is possible. The condition should be suspected in female mice with progressive abdominal distention for a period that exceeds normal gestation. It may also be found when a "male mouse" with nipples is discovered in a female cage. Although a surgical repair of imperforate vagina has been described in the literature, this condition may be inherited, and these animals should not be bred. Female mice may also present with vaginal septa, typically fibrous bands of vertical tissue in the vagina. This condition is more common in mice with a C57BL/6 or BALB/c background. Although mice with vaginal septa are usually fertile, this congenital defect results in a higher incidence of dystocia and poor reproductive outcomes. Mice with vaginal septa should not be bred.

Mild kidney changes, including combinations of tubule degeneration and regeneration, glomerular changes, and interstitial inflammation, can be incidental histopathology findings and tend to increase with age. Severe renal disease can result in protein loss, uremia, and clinical deterioration. Grossly, kidneys can be enlarged or shrunken, with a granular or pitted surface. Histopathology findings can include hypercellular expanded glomeruli and/or shrunken sclerotic glomeruli (glomerulonephritis, glomerulosclerosis), marked tubule dilatation, proteinosis, degeneration, regeneration, and interstitial inflammation and fibrosis.

Chronic urinary obstruction (sometimes known as *mouse urologic syndrome* or *MUS*) can result in enlarged bladder, hydroureter, and hydronephrosis. Hydronephrosis can progress to cause abdominal enlargement. Chronic urinary obstruction leading to uremia can contribute to mortality.

Urogenital tumors

Spontaneous primary urinary tract (kidney and urinary bladder) neoplasms are uncommon in mice. Hematopoietic neoplasms, malignant adrenal neoplasms, or vertebral neoplasms may involve the kidney and urogenital tissues in advanced disease. Spontaneous primary reproductive tract neoplasms (ovarian, uterine, testicular, etc.) are found in aging animals in long-term studies, but usually are incidental findings, not significant contributors to mortality. Endometrial hyperplasia can be common in aging mice. Hematopoietic neoplasms and other invasive neoplasms may involve reproductive organs, and the uterus is a fairly common site for histiocytic sarcoma (Ward 2006).

Common Systemic or Multisystem Conditions

Amyloidosis

Systemic amyloidosis can be a significant cause or contributor to mortality in mice in long-term studies. Mild or early amyloidosis may be an incidental histopathology finding, but clinical deterioration or wasting is characteristic of extensive involvement. Gross lesions may be unremarkable to severe with organomegaly, especially hepato-splenomegaly. Histopathology reveals deposition of acellular fibrillar to amorphous pale eosinophilic material in various tissues. Congo red stain is used to confirm the presence of amyloid. Further characterization as reactive (serum amyloid A, or SAA) type or senile (ApoA) type can be achieved by immunohistochemistry if that should be necessary. Amyloidosis used to be a common problem and cause of death in studies of aging mice, especially of C57BL/6, Swiss, and related mice, but has become less common as colonies have become "cleaner."

Hyalinosis

Hyalinosis refers to eosinophilic cytoplasmic change due to production of eosinophilic chitinase-like protein in epithelia of the glandular stomach, respiratory tract, bile duct, and gallbladder. Usually, this is an incidental finding in susceptible strains. The eosinophilic crystals may be associated with inflammatory conditions. These chitinase-like proteins, especially YM1 crystals, are also found in acidophilic macrophage pneumonia, are produced by neutrophils or macrophages, and can be found in bone marrow and in sites of chronic inflammation.

Conditions related to nutritional status

Nutritional status is mentioned here as it can impact various systems, influence various illness or disease phenotypes, and may contribute positively to, or confound, experimental outcomes. Influences of nutritional status and body condition on experimental parameters should be considered in experimental design and in analysis of results.

While intentional diet restriction in mice may improve life span and reduce age-associated morbidities, inanition or starvation, for only one to a few days, can affect body temperature, size, fat deposits, liver and hepatocyte size, and immune responses. Immune effects, including thymic atrophy and B and T cell suppression, at least in part, are mediated by stress and corticosterone. Some immune responses, such as natural killer cells, macrophages, and granulocytes, increase with short-term inanition. Longer-term inanition also affects multiple systems. Clinically and grossly, mice that do not eat for any reason are expected to be smaller, with lower body condition scores and less fat, and diet-restricted mice have lower fertility. Expected histological changes include less fat and smaller adipocytes (fat cells), smaller hepatocytes, smaller muscle fibers, lymphocyte depletion, and small spleen, thymus, and lymph nodes. Thus, these changes should be interpreted carefully in mice that may not be eating adequately.

Obesity, metabolic syndrome, and type 2 diabetes are important areas of research in mouse models, often intentionally induced or exacerbated by diet in genetically susceptible strains. Obesity is expected to have physiologically significant sequelae, such as alterations in cytokines and chemokines, insulin, and glucose and fat metabolism, and potentially morbid sequelae such as systemic inflammation, hepatic lipidosis or steatosis, hypertension, and/or nephropathy with albuminuria. Standard ad libitum feeding of laboratory rodents leads to obesity in some strains, and age-associated morbidities are increased or accelerated compared with diet-restricted rodents.

Stress-related changes

Sources of stress include disease, diet or water restriction, transportation, social structures, noise, light cycle alteration, temperature extremes, and other changes in environment. Endogenous glucocorticoids, especially corticosterone in rats and mice, are produced during stress and notably affect the immune system. Glucocorticoids induce apoptosis in precursor T and B cells and alter lymphopoiesis. Thymic atrophy can be induced by endogenous or exogenous (administered) glucocorticoids, and apoptosis can be identified by histopathology in lymphoid tissues (thymus, lymph nodes, spleen). In addition, myelopoiesis is stimulated, and blood neutrophil counts are elevated, while blood lymphocyte counts decline in short-term responses to glucocorticoids. Responses vary with mouse strain and are more complex and variable depending on the duration and types of stressors.

common infectious diseases

Given the relatively low prevalence of pathogenic microbes in contemporary colonies compared with a few decades ago, clinically obvious infectious disease conditions are not common in immune-competent mice in reasonably "clean" facilities. Overt enterohepatic disease with diarrhea or severe hepatitis, overt respiratory disease with pneumonia and respiratory noise (chattering) heard in the mouse room, or epizootics with devastating mortality are unusual today. The infectious agents that persist or lurk in contemporary colonies are not as likely to cause substantial morbidity or mortality that raises concern and leads to further investigation. Infections in contemporary colonies are more likely to be detected by surveillance or quarantine testing of animals without overt clinical disease (Pritchett-Corning et al. 2009). However, even inapparent infections modulate the immune system and may interfere with diverse research areas. Obvious or clinical infectious disease problems may signal that:

a. The affected mice are not as immune competent as they are expected to be and may require special handling to protect them from opportunists.

b. Disease is due to a familiar agent that should be identified promptly to protect vulnerable animals, experiments, or personnel.

c. Disease is due to an emerging agent that should be characterized to protect vulnerable animals, experiments, or personnel.

Sources of rodent pathogens exist within and near rodent facilities, as indicated by results from biological materials testing (Chapter 5), surveys of pet shop rodents (Dammann et al. 2011; Roble et al. 2012), and testing of wild rodents near research rodent facilities (Becker et al. 2007; Dyson et al. 2009; Easterbrook et al. 2008; Parker et al. 2009). These risks should be assessed in the development of surveillance, quarantine, and diagnostic testing strategies. Up-to-date information on prevalent murine infectious agents, including diagnosis and treatment recommendations, can be found online (Charles River Laboratories 2020; University of Missouri—Comparative Medicine Program and IDEXX-BioAnalytics 2020).

Significantly, in mice, strains vary in their susceptibilities to infections and in their manifestations of infectious diseases, which can be considered to be disease phenotypes. These variations in disease phenotypes may confound recognition or diagnosis of disease, but also represent opportunities to dissect genetic mechanisms of disease susceptibility and resistance (Sellers et al. 2012).

The agents included below are listed due to either their continued prevalence in modern research facilities or their potentially devastating consequences if found, regardless of prevalence.

Diagnostic Methods

A variety of methods are used to diagnose infectious agents in mice. Different methods or combinations of methods offer advantages for quarantine, surveillance, or diagnostic testing. Optimal testing strategies can depend on the type of facility, microbial exclusion lists of the facility, strain and immune status of the mice, value of the mice, and the cost and time involved.

Pathology: Gross examination of tissues and histopathology are used to detect changes to the host tissues caused by infectious agents. Compared with serology tests and polymerase chain reaction (PCR), which assess only the agents specified, pathology assesses broadly for responses to agents and for other causes or contributors to disease. Organisms such as bacteria and fungi can be identified in tissue sections. Some viruses, such as adenoviruses, herpesviruses, papovaviruses, and poxviruses can leave distinct "footprints" such as intranuclear inclusion bodies. Although not all viruses leave such distinctive footprints, the pathologic changes they cause may suggest a specific agent or agents.

Serologic methods are commonly used to evaluate for the presence of viruses in a colony. *Mycoplasma pulmonis*, *Filobacterium rodentium* (bacteria), *E. cuniculi* (microsporidian), and other agents also may be tested for using serology. Most serologic methods detect antibodies produced by the host against the infectious agent. They do not detect the agent itself. Serologic tests may be positive in the absence of an agent caused by a persistent antibody response to previous, cleared, infection. Serology tests may be negative in the presence of the agent if there is not an effective antibody response, e.g., in immunodeficient mice. However, serologic tests are still the most cost-effective method for large-scale surveillance. Enzyme-linked immunoabsorbent assay (ELISA) is used frequently for primary screening, with immunofluorescent antibody (IFA) tests to confirm findings. Bead-based fluorometric multiplex ELISA (multiplexed fluorometric immunoassay [MFIA]) technologies permit even more tests on smaller samples.

Polymerase chain reaction methods specifically amplify DNA sequences to detect specific sequences of infectious agents, including viruses, bacteria, and eukaryotes. *Reverse transcriptase PCR* (RTPCR) is used to detect RNA sequences in RNA viruses. PCR can be highly specific and sensitive but requires that the correct specimen or tissue(s) be tested for a given agent or agents. PCR methods are becoming increasingly accessible and time saving, and fecal specimens from live animals are proving to be useful for detection of viruses, bacteria, and parasites that are shed from infected animals. The testing laboratory should be consulted regarding its preferred specimens and optimal specimen handling.

Microbial culture is often considered the gold standard for identification of bacteria and fungi. Cultured agents also can be tested for antimicrobial sensitivity. However, isolation of an agent does not necessarily identify it as the cause of the lesion, because opportunistic agents can infect secondarily, and because fastidious agents may be overgrown by faster growing agents or contaminants. Some agents, e.g., obligate intracellular agents, do not grow on artificial media. *Dermatophyte test medium* (DTM) is a specialized agar medium used to selectively cultivate fungi that cause ringworm and indicates presence of their alkaline byproducts with Phenol red pH indicator.

A variety of methods can be used to detect parasites. *Direct microscopic examination* can be used on the pelt or on fur plucks or skin scrapings to detect fur mites, and on various segments of the gastrointestinal tract to identify metazoan parasites. A dissection microscope with a large field and magnification up to 20× facilitates examination. Tape tests involve the application of transparent sticky tape to the pelt or perineum to examine for ectoparasites on the skin or fur, and for *Syphacia* spp. pinworm eggs that are deposited on the perineum. The tape is applied to a glass slide and examined with a microscope. Positive findings are diagnostic, but negative results may not be conclusive. *Fecal flotation* refers to the use of saturated, high-osmolarity solutions to aid in the microscopic detection of protozoal oocysts and nematode eggs in feces. Centrifugation can further concentrate the eggs and oocysts and can increase the sensitivity of the test. A *wet mount* of fresh gastrointestinal contents permits evaluation for characteristic motility patterns of live protozoa. High-magnification phase-contrast microscopy is preferred for these evaluations. PCR of feces or fur or environmental swabs is increasingly being used to detect internal and external parasites, although positive results signify only the presence of genetic material, not necessarily active infectious agents, and so care should be taken in result interpretation.

Viral Agents and Diseases

Multiple surveys throughout the years have detailed the prevalence of various infectious agents. A survey published in 2017 revealed that, compared with information from previous surveys, it appears that prevalence of viral agents is decreasing, while parasitic outbreaks remain the same (Marx et al. 2017).

Viruses that may be found in modern mouse colonies are briefly discussed in this section.

Murine noroviruses (MNV)

Murine noroviruses (MNV) are nonenveloped single-stranded RNA caliciviruses. MNV is the most prevalent virus in research mouse colonies and is transmitted fecal–orally. Its significance in immune-competent mice and the need to eliminate it from research colonies remain controversial, and studies on the effect of MNV on specific models are ongoing. There are usually no clinical signs, but illness including wasting, diarrhea, and death can be seen in severely immunodeficient mice, particularly those with deficiencies in interferon signaling pathways or multiple interferon receptors. Gross pathological findings are uncommon, but histopathology may reveal mild inflammation in the intestine and hyperplasia in the spleen. Inflammatory changes (hepatitis, interstitial pneumonia, pleuritis, peritonitis) may be seen in certain immunodeficient mice. Detection of these viruses is by serology or PCR.

Noroviruses are fairly resistant to routine environmental sanitation using detergents and disinfectants. Primary strategies to eliminate MNV from mice include: (1) test, cull, and decontaminate; and (2) rederivation or fostering into virus-free barriers.

Parvoviruses

Parvoviruses are tiny, nonenveloped single-stranded DNA viruses. *Minute mice virus* (MMV) and *mouse parvovirus* (MPV) are quite common in contemporary mouse colonies. They infect and lyse proliferating cells, such that proliferating immunopoietic or hematopoietic cells, embryonic cells, or cancer cells are damaged or destroyed. Transmission occurs via contact with infected urine or feces. Natural infections are asymptomatic, although there is immune modulation, and there are usually no gross or histopathological findings. Parvoviruses, including MMV, have been identified as contaminants of hybridomas and other transplantable tumors and cell lines. Contaminated feed has been implicated as a source of parvovirus infection, and so irradiation or autoclaving of rodent feed is recommended to prevent introduction of parvoviruses via the feed (Adams 2019).

Serologic testing for MMV and MPV can be frustrated by slow or weak seroconversion, as well as mouse strain–related variations in seroconversion. PCR of mesenteric lymph nodes or feces may be useful in detecting active infections with viral shedding.

These hardy viruses persist in the environment, resist many decontaminants, and are highly infectious. Biological materials, wild mice, contaminated food or bedding, and other fomites are potential sources of infection. Rederivation or fostering into parvovirus-free barriers is the most reliable approach for elimination of these viruses from infected colonies.

Murine chapparvovirus (MuCPV), aka mouse kidney parvovirus (MKPV), is antigenically distinct from MVM and MPV and is moderately prevalent in modern laboratory mouse colonies. Transmission is via urine and feces, and the disease does not tend to cause clinical signs in immune-competent mice. In immune-deficient mice, chronic renal failure may result, with the most severe effects seen in severely immunocompromised strains such as *Rag* knockout, SCID, and NSG mice. Histopathology reveals inclusion body nephropathy, with intranuclear inclusion bodies in renal tubular epithelial cells, tubular degeneration, and necrosis. Diagnosis is made via PCR of kidney, feces, or environmental samples. Testing of dirty bedding sentinel animals does not reliably detect the virus, and it is not detected by currently available serological and PCR tests for MPV or MVM. Because the agent has recently been identified, the full pathobiology and significance of the virus are not known.

Mouse hepatitis virus (MHV)

Coronaviruses, including mouse hepatitis virus (MHV), are enveloped single-stranded RNA viruses. Disease in mice varies with MHV strain, dose, and route of administration, as well as with mouse strain, age, and immune status. MHV strains have been classified as: (1) primarily *enteric* or *enterotropic* strains, with infection and lesions usually restricted to the gut of young animals; and (2) *respiratory* or *polytropic* strains that infect the respiratory tract initially, then spread to other tissues. Historically, MHV was prevalent in research colonies, and enterotropic strains were an important cause of epizootics (outbreaks) of infant diarrhea and mortality; hence the acronym *LIVIM* for lethal intestinal virus of infant mice. Transmission occurs through contact with infected mice, including wild mice, fomites, airborne particles, and biological materials. Transplacental infection has been documented experimentally. Because respiratory or polytropic strains infect various tissues, they are more likely to be encountered as contaminants in biological materials.

Enterotropic MHV strains infect the intestinal epithelium in all ages of mice. Naïve suckling mice are susceptible to diarrheal disease and mortality. In contrast to rotaviral diarrhea, infected neonates stop nursing and may die (LIVIM). Older, immune-competent mice are infected but usually do not develop disease. Enterotropic MHVs are highly contagious, with susceptible or immunodeficient mice shedding abundant virus in feces for long periods. Enzootic infection of a colony is usually subclinical, perpetuated by breeding mice that transmit the virus, along with protective immunity, to offspring. There are often no gross pathological findings in infected mice, but typhlocolitis or hepatic necrosis may be seen in susceptible mice. Syncytial cells may be detected in the ascending colon.

Polytropic or respiratory strains infect the respiratory tract initially, then disseminate. Clinical disease in natural infections of competent mice is unusual, but in immunodeficient mice there can be progressive wasting and necrotizing liver disease, typhlocolitis, and/or encephalitis. The classic gross lesion in disseminated MHV is the presence of white foci on the liver. MHV is typically detected using multiplex or ELISA technologies, with confirmation by IFA. PCR of feces may be useful in actively infected animals that are shedding virus. Histopathology findings of hepatitis, necrosis or typhlocolitis, or syncytia should lead to further testing for MHV.

MHV is relatively fragile in the environment and susceptible to common sanitation procedures. Primary strategies to eliminate MHV include: (1) test, cull, and decontaminate; or (2) rederivation or fostering into MHV-free barriers.

Theiler's mouse encephalomyelitis virus (TMEV)

Murine encephalomyelitis virus (MEV, TMEV) is a nonenveloped RNA virus. Transmission is via the fecal–oral route, or through infection of biological materials. Natural infection usually is subclinical, but fatal encephalitis may be seen in infection with virulent strains. Experimental inoculation results in demyelinating lesions, particularly in SJL/J mice. Diagnosis is via serology or PCR.

Mouse rotavirus (MRV, EDIM)

MRV is a mouse-specific, nonenveloped double-stranded RNA virus. MRV causes diarrhea in young animals and has been used to model human rotavirus diarrhea. Transmission is fecal–oral.

Multiple strains of MRV have been isolated and identified as causes of epizootic diarrhea, typically yellow, in infant mice (originally epizootic diarrhea of infant mice, or EDIM). Pups continue to nurse and have milk in their stomachs, evident through their thin skin. They usually survive but may be runted. Histologically, vacuolation of apical villus epithelium in small intestine is transient and may be difficult to distinguish from normally lipid-laden villus epithelium in nursing neonates. Cytoplasmic inclusion bodies in enterocytes have been described; special stains may be required to visualize them. Older mice can be infected but do not develop disease. Even immunocompromised mice are susceptible to disease only up to about 15 days of age.

MRV is not common in modern mouse colonies, but it should be considered when there is neonatal diarrhea, runting, high morbidity, low mortality, and absence of clinical disease in older mice. Diagnosis is typically done through serologic testing using multiplex or ELISA technologies, with confirmation by IFA. RTPCR of feces may be useful in actively infected animals that are shedding virus.

MRV is fairly resistant to elimination by sanitation using detergents and disinfectants. Primary strategies to eliminate MRV include: (1) test, cull, and decontaminate; or (2) rederivation or fostering into virus-free barriers. Cessation of breeding to break the cycle of continuous fecal–oral transmission in breeding situations has been reported to be effective (Held et al. 2011), but should be considered only under certain circumstances (see section titled "Treating Disease on a Colony Basis").

Murine astrovirus

Murine astrovirus (MuAstV) is a nonenveloped single-stranded RNA virus that was first identified in a population of nude mice with diarrhea, although the virus was not definitively determined to be the cause of the diarrhea (Kjekdsberg and Hem 1985). While the virus appears to be widespread in modern research colonies (Ng et al. 2013), immune-competent mice do not display clinical signs (Compton et al. 2017), and impacts to research remain unknown.

Mouse adenoviruses (MAV 1 and 2)

Adenoviruses are relatively large, nonenveloped double-stranded DNA viruses. These viruses are uncommon in contemporary research colonies, and natural infections do not cause clinical disease.

Lymphocytic choriomeningitis virus (LCMV)

LCMV is a small, enveloped single-stranded RNA virus. LCMV has been significant in laboratory mice as a contaminant of biological materials and because of its zoonotic potential. LCMV can be transmitted through contact with saliva, nasal secretions, or urine. LCMV can infect various species including mice, rats, hamsters, and humans. The virus in mice is usually subclinical. Because of its zoonotic potential, if LCMV is identified in a mouse colony, all colony mice should be euthanized.

Lactate dehydrogenase–elevating virus (LDV, LDEV)

LDV is a small, enveloped single-stranded RNA virus that is primarily transmitted in the laboratory setting through infected biological materials, and so testing of biological materials is important for prevention. This virus usually causes no clinical disease, although flaccid paralysis and poliomyelitis have been reported in immunosuppressed C58 and AKR mice, and in $Prkdc^{scid}$ mice that were co-infected with an endogenous murine leukemia virus. Diagnosis is based upon elevated LDH activity in serum or plasma, i.e., levels in excess of three to six times the reference range.

Hantaviruses (Han)

Hantaviruses are enveloped single-stranded RNA viruses. They are not likely to be encountered as natural infections in contemporary mouse colonies, but seropositivity has been reported in laboratory mice (Won et al. 2006). They are of concern primarily because of their zoonotic potential. Infected animals generally do not display clinical signs. Diagnosis is through serology or PCR.

Herpesviruses

Herpesviruses are large, enveloped double-stranded DNA viruses. There are two herpesviruses that may infect mice (mouse cytomegalovirus and mouse thymic virus), although natural infections with these viruses are uncommon in contemporary mouse colonies.

Ectromelia virus (mousepox)

Ectromelia virus is a large, enveloped double-stranded DNA virus and is the agent of mouse pox. Historically, this agent caused epizootics in susceptible mice, but it is not often seen in modern mouse colonies.

Transmission occurs via direct contact or fomites. Highly susceptible strains, such as BALB/c or DBA, can die quickly before development of characteristic lesions or viral shedding. Resistant strains such as C57BL/6 can harbor the agent subclinically but serve as a source of infection of other animals. Intermediately susceptible animals may display nonspecific signs such as ruffled fur and hunched posture, along with the characteristic cutaneous pustules and ulceration of the muzzle and extremities. *Ectromelia* refers to the partial amputation of the limbs that can occur in mice that survive. Diagnosis is done through serologic testing, characteristic histopathology lesions in liver and spleen, and/or PCR of skin lesions. Screening of biological materials is important to prevent introduction of the virus via cell lines.

Papovaviruses

Papovaviruses are enveloped double-stranded DNA viruses, three of which—mouse papillomavirus, K virus, and mouse polyomavirus—have been recognized as mouse pathogens, although they are vanishingly rare in modern mouse facilities.

Mouse papillomavirus has been identified in proliferative skin lesions and papillomas in nude mice and is used as a model for human papillomavirus (HPV) (Held et al. 2011).

K virus (murine pneumotropic virus) was discovered originally as a contaminant of transplantable mouse tumors. Transmission is via the fecal–oral route. Natural infection, which is highly unlikely in a modern mouse colony, tends to be subclinical and persistent, although infection of newborns can result in pneumonia.

Mouse polyomavirus (murine polyomavirus) causes subclinical infections in immunocompetent adult mice. Natural infections with this virus are rare in modern mouse colonies. Experimental infections of neonates or athymic nude mice can lead to tumors in multiple tissues (poly + oma, meaning many tumors). Nude mice may develop multifocal necrosis and inflammation with intranuclear inclusion bodies, possibly paralysis due to vertebral tumors, or demyelination. The characteristic tumor resulting from experimental infections of neonates is a pleomorphic salivary gland tumor, called a *myoepithelioma*, with concurrent inflammation that is unusual in spontaneous salivary gland tumors. Tumors also can be induced in other tissues, especially in susceptible strains.

Papovaviruses can be diagnosed via serology or PCR.

Paramyxoviruses

Paramyxoviruses are enveloped single-stranded RNA viruses. Sendai virus and pneumonia virus of mice are the significant paramyxoviruses in mice. Their prevalence in laboratory mice has diminished significantly in the last three decades.

Pneumonia virus of mice (PVM) infects laboratory mice, rats, and hamsters. Transmission is via aerosol and contact with respiratory secretions. Natural infection usually is subclinical in immunocompetent mice, but immune compromised mice may develop chronic wasting disease with progressive interstitial pneumonia.

Sendai virus is uncommon since the advent of effective isolator caging. Transmission is via aerosol or contact with respiratory secretions. Sendai virus is one of the few mouse pathogens that can cause clinical signs in immune-competent animals. These signs include dyspnea, chattering (respiratory noise), and high mortality in susceptible mice. Susceptibility is strain-dependent, with DBA/2 mice very susceptible, and C57BL/6 mice resistant.

Diagnosis of paramyxoviruses is via serology or PCR.

Reovirus 3 (REO3)

Reovirus 3 (REO3) is a nonenveloped RNA virus. Transmission is via the fecal–oral route, direct contact, or fomites. Natural infections are uncommon in contemporary mouse colonies. Usually, there are no clinical signs associated with infection, but stunted growth, diarrhea, jaundice, and oily hair coat, sometimes called OHE (oily hair effect), have been associated with infection in susceptible young mice. Diagnosis is via serology or PCR.

Bacterial Agents and Diseases

Bordetella species

Bordetella is a Gram-negative small rod bacterium. In mice, these bacteria usually are opportunist agents. There are recent reports of respiratory disease associated with *B. hinzii*, which shares many features with *B. avium*, and possible interference with pulmonary models (in the absence of overt signs of respiratory disease) associated with *B. pseudohinzii* (Clark et al. 2016, 2017). *Bordetella bronchiseptica* can be highly pathogenic in guinea pigs and is a common agent in rabbits, but it is not commonly found in mice.

Filobacterium rodentium (formerly cilia-associated respiratory bacillus, CAR bacillus, CARB)

A Gram-negative small filamentous rod, *Filobacterium rodentium* was recently classified, after being referred to as cilia-associated respiratory bacillus, or CAR bacillus, for over 35 years (Ike 2016). In mice and rats, it is recognized as an important co-pathogen with *Mycoplasma pulmonis*, but it can cause pneumonia in the absence of *M. pulmonis*. Histopathology findings include bronchopneumonia, with abnormally clumped cilia in the respiratory epithelium. Because of its unique culture requirements, diagnosis is best made through serology screening with subsequent PCR for confirmation.

Citrobacter rodentium

C. rodentium is a Gram-negative small rod and the etiologic agent of transmissible murine colonic hyperplasia. Clinical and gross pathological signs of infection include diarrhea, rectal prolapse (Figure 4.11), and grossly thickened large bowel. Suckling mice are especially susceptible to clinical disease caused by *C. rodentium*.

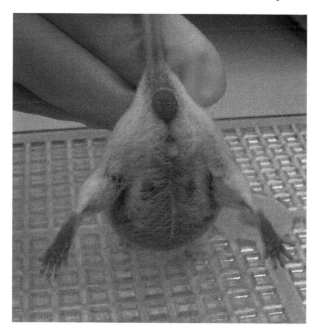

Figure 4.11 Rectal prolapse in a mouse. This condition has been associated with infection by *Helicobacter* spp., *Citrobacter rodentium*, pinworms, or other conditions that cause diarrhea or straining.

Competent mice clear the agent rapidly, and so it can be difficult to detect, even in affected tissues. Although common in the past, natural infection is unlikely in contemporary colonies. Diagnosis is through culture.

Clostridium piliforme

Clostridium piliforme is a Gram-negative rod and the agent of Tyzzer's disease. Although uncommon today, it was once recognized as an important pathogen of mice. Infections can be subclinical, but mortality may be common with poor husbandry and immunosuppression. Transmission is fecal–oral or through ingestion of spores from the environment. Clinical signs are typically seen in recently weaned animals and can include watery diarrhea, lethargy, ruffled hair coat, and sudden death. Important differential diagnoses for diarrhea in young mice include EDIM and MHV infection. The most consistent gross finding in mice is multiple pale foci of necrosis in the liver (multifocal necrosis). PCR on fecal samples can help identify actively shedding animals.

Corynebacterium bovis

Corynebacterium bovis is a Gram-positive small pleomorphic rod that causes "scaly skin disease," a scaly dermatitis, primarily in nude and immunodeficient mice. Most mice survive the acute infection, but acanthosis remains, and shedding of the organisms continues, leading to enzootic infections in which animals remain persistently infected, but show clinical signs less frequently. Transmission is via direct contact or fomites. Diagnosis is through culture or PCR of skin, oropharynx, or contaminated environment. Eradication can be difficult and typically requires rederivation and aggressive environmental decontamination. This infection can be associated with pup mortality in breeding colonies of immunodeficient mice, although metaphylactic antibiotic administration may prevent transmission to offspring (Pearson et al. 2020). Histopathology findings of acanthosis and hyperkeratosis, with intracorneal and intrafollicular colonies of Gram-positive small pleomorphic bacteria, are characteristic. PCR from skin swabs or flakes can be useful for rapid detection.

Helicobacter

Helicobacter species are Gram-negative curved or spiral bacteria. The prevalence of *Helicobacter* spp. in research colonies is quite high, although the species specifically associated with clinical disease in mice are *H. bilis* and *H. hepaticus*. Transmission is by the fecal–oral route. The primary clinical sign associated with *Helicobacter* infection is rectal prolapse secondary to typhlitis or typhlocolitis. Diarrhea may also be seen. Some *Helicobacter* spp. have been identified as contributors to liver tumors in susceptible strains, and *Helicobacter* may interfere with reproductive success (Bracken et al. 2017). Fecal PCR is the most common and practical test method for diagnosis of *Helicobacter*. Establishment of a *Helicobacter*-free colony typically requires aseptic hysterectomy or embryo transfer rederivation.

Mycoplasma

Mycoplasmas are the smallest and simplest bacteria. They lack a rigid cell wall and so are pleomorphic. They are Gram negative but stain poorly. *Mycoplasma pulmonis*, the agent of murine respiratory mycoplasmosis, was once one of the most important pathogens of laboratory rats and mice but has been largely eliminated from contemporary research colonies.

Mycoplasma pulmonis is primarily a pathogen of the respiratory tract, but also can cause disease in the reproductive tract (genital mycoplasmosis) and arthritis. Transmission occurs via direct contact, aerosol, and transplacentally. Clinical signs in susceptible, chronically infected animals include chattering, dyspnea, weight loss, hunched posture, and lethargy. Immunodeficient mice are particularly susceptible to pneumonia and death, and they may develop severe arthritis following infection. Diagnosis is via serology or PCR of lung lavage fluid. *Filobacterium rodentium* is a frequent co-pathogen.

Rodentibacter pneumotropicus and *R. heylii* (formerly *Pasteurella pneumotropica* type Jawetz or Heyl)

Gram-negative coccobacilli, *Pasteurella* and related species, can be commensal agents or pathogens in diverse host species. The taxonomy of *Pasteurella* species changes regularly, with the two subtypes of *P. pneumotropica* reclassified as *Rodentibacter pneumotropicus* and *R. heylii* as of 2018 (Adhikary et al. 2017).

Rodentibacter spp. have been implicated in various clinical syndromes, including conjunctivitis, infections of the respiratory and reproductive tracts, otitis, and subcutaneous abscesses, especially in immunodeficient mice. Transmission is via direct contact with infected animals or their secretions. Detection is through PCR or culture of abscesses or usual colonization sites including nasopharynx, vagina, and intestines. Due to the prevalence of these organisms as commensal agents, eradication through antibiotic treatment may be difficult, although successful treatment with enrofloxacin in drinking water has been reported (Goelz et al. 1996; Matsumiya and Lavoie 2003; Towne et al. 2014). Rederivation through embryo transfer may be necessary for elimination.

Pseudomonas aeruginosa

P. aeruginosa is an aerobic, Gram-negative rod, frequently found in the environment. Exposure of laboratory mice to *P. aeruginosa* is typically through watering systems, and the organism remains one of the primary targets of water treatment strategies in laboratory animal facilities. Disease caused by *P. aeruginosa* is primarily seen in animals made neutropenic through irradiation or chemotherapy and includes bacteremia and sepsis. Diagnosis is via bacterial culture.

Staphylococcus

Gram-positive plump cocci, Staphylococci are commonly isolated from healthy animals, but also are capable of causing disease, especially via contaminated wounds or in immune compromised individuals.

Staphylococcus aureus is a fairly common resident of the skin of many animals, including mice. Entry of the organism into the body is via breaks in normal barriers (e.g., wounds or ulcerative dermatitis), often resulting in abscessation. Clinical signs of S. aureus in mice include soft-tissue swellings due to furunculosis or facial abscesses (Figure 4.12), or preputial and clitoral gland abscesses, and may include sepsis and death in susceptible mice. Diagnosis is via culture or histopathology. The classic S. aureus histopathology lesion is a dense mass of cocci surrounded by Splendore–Hoeppli material, strongly eosinophilic amorphous material with a radiating configuration, thought to be due to deposition of antigen–antibody complexes and debris from host inflammatory cells (Gopinath 2018). This is also known as botryomycosis.

Depending on the agent, the extent of clinical signs, and the number and value of mice affected, systemic antibiotic treatment may be warranted for bacterial infections in mice. There are a variety of antibiotics that can be administered via oral or parenteral routes. Care should be taken when treating an active infection with antibiotics administered via drinking water, as plasma concentrations may not reach therapeutic levels when antibiotics are administered

Figure 4.12 Mouse, face and paw demonstrating botryomycosis. *Staphylococcus aureus* was isolated. The mouse was immunodeficient.

in this manner (Marx et al. 2014). To ensure dosing and plasma levels adequate for elimination of infection, parenteral administration is recommended.

Fungal Agents and Diseases

Pneumocystis murina

Pneumocystis murina is classified as a fungus, although it has many protozoan characteristics. In immunodeficient animals, *P. murina* causes chronic progressive pneumonia. Clinical signs include wasting, rough hair coat, dyspnea, cyanosis, and death. Gross findings include pale gray or red areas of lung consolidation and lungs that do not deflate. Diagnosis is via histopathology or PCR of nasal swabs, lung tissue, or deep bronchoalveolar lavage. Several antibiotics have been used to control infection in immunodeficient mice, but they do not eliminate the agent, and disease can be expected to recur after treatment is withdrawn.

Intestinal protozoa

There are several intestinal protozoal species classified as "pathogenic" in laboratory mice, although infection is rare and disease is subclinical in immune-competent animals: *Cryptosporidium muris*,

Cryptosporidium parvum, Eimeria spp., *Entamoeba muris* (Figure 4.13), *Giardia muris,* and *Spironucleus muris* (Figure 4.14). Transmission tends to be fecal–oral. Immune compromised animals may display non-specific signs such as ruffled fur, hunched posture, and rough hair coat. Diarrhea or soft stool may also be seen. Diagnosis is

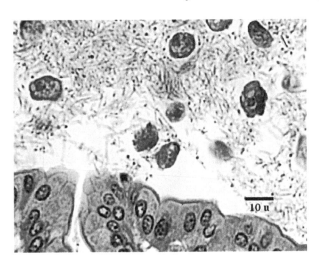

Figure 4.13 Colon and colon content, *Entamoeba muris* in the lumen of the colon.

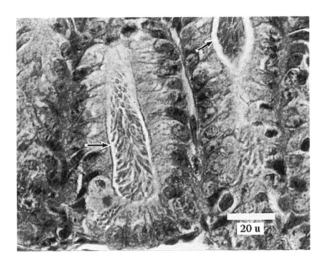

Figure 4.14 Small intestine, demonstrating numerous small elongate *Spironucleus muris* trophozoites (arrows) within crypts.

typically via fecal examination, although histopathological evalua-tions of portions of the GI tract may be necessary to identify some species.

Parasitic Agents and Diseases

Arthropods

Fur mites and mesostigmatid mites continue to be annoying and expensive problems in contemporary mouse colonies. Follicle-dwelling mites and lice are uncommon in contemporary mouse colonies and may indicate exposure to wild rodents or to pet store rodents. Nonparasitic arthropods in the mouse environment, such as psocids, or booklice, are pests that should be distinguished from parasites.

Fur mites

Fur mites remain relatively prevalent in laboratory mouse colonies. Three species of fur mite are associated with infestation in labora-tory mice. *Myobia musculi* (Figure 4.15) and *Myocoptes musculinus*

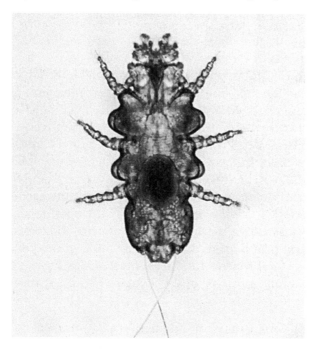

Figure 4.15 Fur mite: *Myobia musculi.*

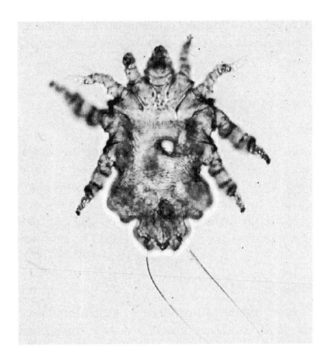

Figure 4.16 Fur mite: *Myocoptes musculinus.*

(Figure 4.16) are the most commonly identified fur mites of mice. *Radfordia affinis* resembles *Myobia musculi* and is less frequently found. Mixed infestations are common. *Myobia* and *Radfordia* tend to be found near the face and neck, while *Myocoptes* tend to be located near the rump or flank. Mites are transmitted via direct contact, and their life cycles are direct, with all stages (egg, nymph, and adult) living on the host. Because of this, dirty bedding sentinels are not useful for detecting infection. In most cases, infections are subclinical, although heavy infestations or infection of immune compromised mice can result in pruritis, alopecia, ulcerative dermatitis, and secondary bacterial infections. PCR of skin or cage swab is the most reliable detection method. Less sensitive methods include microscopic examination of fur plucks and postmortem direct examination of the pelt. Treatment with topical or oral ivermectin-family insecticides have been effective. Table 4.3 summarizes some of the features of mouse fur mites.

Demodex

Demodex musculi is a follicular mite found to infect laboratory mice. Reports of infection are rare, but in truth infection may be

TABLE 4.3: FEATURES OF COMMON FUR MITES IN MICE[a]

	M. musculi	*R. affinis*	*M. musculinus*
Body shape	Elongate body with bulges between the legs	Elongate body with bulges between the legs	Rounded oval
Distinguishing features	1st pair of legs (tarsi) are specialized for hair clasping; 2nd pair of legs has a single empodial claw	1st pair of legs (tarsi) are specialized for hair clasping; 2nd pair of legs has 2 unequal claws	3rd and 4th pairs of legs (tarsi) are short thick and specialized for hair clasping
Adult size	300–500 μm long	300–500 μm long	Females oval 130 × 350 μm Males rounded <200 μm diameter
Eggs	Oval up to 250 μm long; base of hair shaft	Probably similar to *M. musculi*	Oval up to 450 μm long; distal hair shaft
Diet	Feeds on interstitial fluids	Probably similar to *M. musculi*	Feeds on superficial keratin layer
Site on mouse	Deep among hairs especially of dorsal head, neck, shoulders, and flank		Among the hairs especially of inguinal skin and ventrum

[a] See Baker, D. G., *Flynn's Parasites of Laboratory Animals*. 2nd ed., Blackwell Publishing, 2007, for further details.

more prevalent yet go undetected (Nashat et al. 2018). *Demodex* are commonly regarded as commensal organisms but can cause dermatologic disease in immune compromised mice (Nashat et al. 2017; Smith et al. 2016). Microscopic examination of deep skin scrapes and fur plucks of the interscapular region and caudal ventrum, as well as PCR of feces or environmental samples, has been found to be effective for diagnosis (Nashat et al. 2018b). Eradication can be difficult, but extended topical moxidectin–imidacloprid administration proved successful in one outbreak (Nashat et al. 2018).

Helminths

Helminth parasites in mice include cestodes and nematodes. Of these, only pinworms remain common in contemporary mouse colonies.

Nematodes

Aspiculuris tetraptera and *Syphacia muris* are the pinworms of mice. Both have a direct life cycle, and transmission is via ingestion of infectious embryonated eggs. In immune-competent animals, infection is usually subclinical. In immune suppressed animals or those with comorbidities, rectal prolapse, weight loss, and poor hair coat may be seen. PCR on feces is a reliable diagnostic test for both species of pinworm. Direct exam of cecal or colonic contents will identify adults of *Syphacia* or *Aspiculuris*, respectively, although euthanasia is required. Fecal flotation will identify eggs of either species. *Syphacia* deposits eggs in the perianal region, while *Aspiculuris* releases eggs in the colon, making the perianal tape test useful only for detection of *Syphacia* ova. The two species have distinctly different egg shapes, with *Syphacia* eggs being asymmetric and flattened on one side (Figure 4.17), while *Aspiculuris* eggs are bilaterally symmetrical (Figure 4.18). Treatment generally involves an alternating-week regimen of fenbendazole-containing feed. Pinworm eggs are very light and can become airborne, resulting in widespread environmental contamination. The eggs survive for variable periods in the environment; they resist desiccation and common disinfectants but are susceptible to high temperatures. This can make environmental disinfection difficult and can lead to reinfection after presumed eradication (Table 4.4).

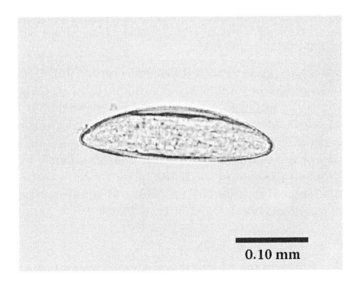

0.10 mm

Figure 4.17 A *Syphacia* egg found during examination of a fecal flotation sample. Note the flattening of one side. These eggs are usually found by applying transparent tape to the perineum of the mouse, where they are deposited.

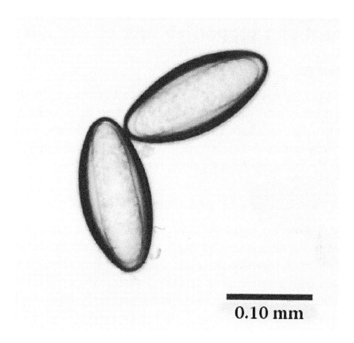

0.10 mm

Figure 4.18 Two *Aspiculuris* eggs from a fecal flotation. Note the bilateral symmetry compared with the *Syphacia* egg (Figure 4.17).

TABLE **4.4:** COMPARISON OF FEATURES OF *SYPHACIA OBVELATA* AND *ASPICULURIS TETRAPTERA*

	Aspiculuris tetraptera	*Syphacia obvelata*
Prepatent period (egg to egg)	21–25 days	11–15 days
Infectious form	Embryonated eggs	Embryonated eggs
Larvae hatch from eggs	Cecum	Cecum
Adult size	2–4 mm long, <200 µm wide	<6 mm long, <200 µm wide
Adult location	Colon lumen	Cecum lumen, rectum
Number of eggs	17 per female per day, intermittently	350 per female all at once
Eggs	Not embryonated; shed in feces; embryonate in 5–8 days	Not embryonated; deposited on perineum; embryonate within 24 h
	41 µm wide, 90 µm long; symmetric and unembryonated in fresh feces	36 µm wide × 134 µm long; crescent shaped, flattened on 1 side, with pointed ends

treatment and supportive care of sick mice

Drug Dosages

Treatment of sick mice should be implemented under the direction of a qualified veterinarian, following appropriate diagnostic measures. Many institutions publish formularies with drug doses recommended by the institutional veterinarian(s). These references should be consulted for the most up-to-date dose recommendations.

General Treatment of Open Skin Lesions

Mice may develop open skin lesions for a variety of reasons, including fighting and self-trauma resulting from infestation with ectoparasites. It is important to identify the most likely cause of the lesions so that appropriate treatment can be initiated. If the lesion is believed to be due to fighting, affected mice should be promptly separated from the aggressor and monitored carefully for ongoing fighting. All mice with fight wounds should be evaluated to ensure that they can urinate, even when the skin lesions do not appear severe, as the genitals are a common target for aggressors, and urinary blockage associated with tissue damage can be life-threatening. Treatment for deep or extensive lesions should include careful flushing with sterile saline or dilute antiseptic solution and the provision of analgesia. Care should be taken when instituting topical treatments such as antibiotic ointment for skin lesions as mice may be more likely to lick, bite, or scratch at the area to which the treatment is applied, possibly introducing additional contamination or tissue damage. In the case of ulcerative dermatitis, the literature abounds with proposed treatments, from dietary supplements to topical administration of dilute bleach (Hampton et al. 2015; Michaud et al. 2016; Whiteside et al. 2020). Nail trimming has been found to be an effective treatment that is less likely to interfere with research results than treatments involving pharmacologic agents (Alvarado et al. 2016). Unfortunately, no treatment has been found to be universally effective. In general, mice with severe skin lesions or lesions that do not respond to treatment should be euthanized.

General Treatment of Weak Mice

Many disease processes in mice result in a similar outcome: a weak mouse. Typically, these mice have a hunched appearance, are lethargic and unwilling to ambulate, have an unkempt appearance, may be thin, and are sometimes hypothermic. In many cases, euthanasia may be the most humane and reasonable option. However, regardless of the specific disease, some general steps can be taken to provide care to such animals.

Provision of supplemental fluids

A sick mouse is one that may not be drinking and is dehydrated. In general, one can assess the hydration status of a mouse by gently pulling a tent of skin up on the back of the mouse and evaluating how readily the skin returns to the original position. In a well-hydrated animal, this will take, perhaps, 1 second. In a dehydrated mouse, the skin may not return to the original position for 2 or 3 seconds or may not readily return to that position at all. In such cases, the animal should receive supplemental fluids. The exact volume of fluid to be given is best based upon the extent of dehydration, though that is difficult to determine in mice. Instead, a general volume of 0.5–1.0 ml is a good starting point, and this can be supplemented if needed. Fluids can be given subcutaneously in non-emergent situations, while intraperitoneal administration is preferred for extreme cases due to more rapid absorption. Oral gavage can be stressful, and this administration route should be avoided, if possible, in animals whose health is already compromised. Ideally, fluids are first warmed to approximate normal body temperature of the mouse. Appropriate fluids include sterile normal (0.9%) saline or lactated Ringer's solution. A 5% dextrose solution might provide additional energy support to dehydrated animals that are also displaying signs of weakness. Steps should be taken to first ensure that the dextrose will not interfere with experimental objectives.

Provision of food

As with dehydration, mice may not have sufficient nutritional intake if they are inappetent or unable to access food due to pain, weakness, or mobility issues. In these cases, nutritional supplementation should be provided in addition to their normal feed and location. For example, if they are unable to ambulate to or reach the normal location of their feed, pellets can be placed within a dish on the cage floor. Moistening the pellets with water from the animal's usual water source can create a mash that is more easily consumed and that provides supplemental oral fluids as well. Commercially available gel diet products may be more palatable than standard chow and can provide additional nutritional and hydration support. In cases where mice appear lethargic, oral administration of several drops of 50% dextrose can be considered.

Provision of supplemental heat

Sick mice can easily become hypothermic due to their relatively great body surface area to body mass ratio. As a result, it is advisable to provide such animals with supplemental heat. One basic approach is to maintain the animal with cagemates and adequate nesting material, since mice tend to build a nest and sleep there together, thus keeping one another warm. Of course, if the animals begin to fight or attack the sick animal, they should be removed. Likewise, if isolation of the sick animal is chosen for control of suspected infectious disease, this approach would not be practical. Provision of nesting material and abundant bedding may also help mice avoid hypothermia. Alternatively, the cage can be placed upon a heat source such as a circulating water heating pad or a temperature-controlled kennel heater. It is important to place only a portion of the cage over the heating source so that mice can seek a cooler area if desired. Although heat lamps can be effectively used as supplemental heat when placed over a cage of mice, the risk of overheating outweighs the potential benefit, and they are therefore not recommended.

Treatment of Dystocia

Difficulty associated with the birth of pups is a commonly encountered problem in breeding colonies of mice. Parity, obesity, and age at first litter are all associated with dystocia. Sometimes, this results from a fetus being lodged in the birth canal, while other times it is caused by a seeming lack of sufficient uterine contraction. The duration of labor in mice varies greatly with strain and genotype, but typically ranges from 30 minutes to several hours, with the onset in the dark cycle. In some cases, labor may last longer, and the presence of normal, well-cared-for pups and a clinically normal dam all suggest normal labor. In contrast, dystocia is suspected when the dam appears to be in distress, when labor has yielded only a few pups, and when the dam fails to tend to the newborn pups that may lie scattered outside of the nest. Dystocia can result from a variety of causes and often requires (futile) veterinary intervention.

An important initial step is to examine the vaginal canal for fetuses that are stuck. If present, it is sometimes possible to manually remove the fetus. With a cotton-tipped applicator, a small amount of lubricant can be applied to the fetus and an attempt made at gentle removal.

Often in dystocia, the dam is dehydrated. This can further compound the difficulty of giving birth, and thus provision of supplemental fluids is important. Both the fluid volume and the electrolytes offered by some solutions provide benefit. As described previously in the section on provision of supplemental fluids, lactated Ringer's solution or normal saline are of value. Various pharmacological treatments have been described, but none has been published in the literature, and so the authors are unable to recommend a specific pharmacologic cocktail.

Often, mice in dystocia cannot be rescued. If steps taken as described do not result in delivery of pups within 1–2 h, further efforts are likely to be futile, and the dam should be euthanized, and any surviving pups euthanized or transferred to a foster dam. If dystocia is recognized early enough and the pups are valuable, the dam can be euthanized, and a hysterectomy can be performed with subsequent fostering. Oftentimes, however, by the time signs appear, the unborn pups are unlikely to survive. If the dystocia is resolved, the dam should be carefully palpated to ensure there are no retained pups. If retained pups are identified and are not passed within several hours, the dam should be euthanized or, if her pups are valuable and a foster is not available, she can be treated with parenteral antibiotics to prevent infection until her pups are weaned and she can be euthanized.

clinical endpoints

Mice used in research can be particularly valuable, either because of the specific experiment in which they are being used or because they have a valued genotype. For this reason, the decision to euthanize a sick animal is difficult when treatment has failed or is impractical. Defining specific points at which an animal is euthanized when clinical decline due to spontaneous or induced disease develops is useful in this regard. In cases of disease induced due to genotype or experimental manipulation, it is useful to define such endpoints when the institution's oversight body (e.g., IACUC or other animal care and use oversight body) is reviewing proposed research protocols. In this way, investigators, animal care personnel, and the IACUC can all reach consensus on specific endpoints before the study has begun, and animals reaching those endpoints can be expeditiously euthanized.

The specific endpoints applied to research animals may vary from study to study. For example, if justified by the investigator for scientific purposes, later endpoints, even up to death of the animal, may be approved by the IACUC. Nonetheless, some general guidelines will be offered here for consideration in developing clinical endpoint criteria. Objective measurements and a scoring system that reflects varying degrees of severity for each criterion, with specific actions based on the total score, are useful tools for an objective determination of humane endpoints.

In general, endpoints based on clinical criteria should strive to minimize animal pain and distress to only that which is unavoidable while meeting the scientific needs of the study. For example, unless scientifically justified, using the moribund state as an endpoint might be inappropriate since mice could have already progressed through stages of decline that involved significant pain and distress; it would be desirable to instead identify endpoints prior to the moribund state. Typically, other clinical signs can be used as a proxy to identify mice that are likely to progress to moribundity, so that they can be humanely euthanized prior to that point. It is important to note that, as prey animals, mice are likely to hide obvious signs of pain and distress until a certain threshold is reached. As such, mice that are exhibiting signs may be significantly affected. In mice, signs of pain or distress might include reluctance to move and lethargy; failure to groom and resulting unkempt appearance of the hair coat; hunched appearance, with legs and abdomen tucked and head down; lack of appetite and thin body condition; loss of nest building behavior; cannibalization of neonates; and, in some cases, vocalization. Lack of vocalization should not be taken as lack of pain or distress, because mice typically vocalize at frequencies that are inaudible to humans. A system of scoring body condition in mice as a means to identify endpoints is also useful (Ullman-Cullere and Foltz 1999). Pain in mice can also be evaluated by mouse facial expression, with orbital tightening (squinting), nose bulging, cheek bulging, drawing of the ears back against the head, and certain movements of the whiskers (collectively referred to as the Mouse Grimace Scale) all believed to be indicators of pain (Langford et al. 2010). Effective use of the Mouse Grimace Scale requires training; many components may be difficult to evaluate without the use of high-speed cameras and may be affected by prior use of anesthetic agents or sedatives (Tappe-Theodor et al. 2019).

Broad clinical considerations in determining endpoints may include ensuring that animals can easily access feed and water; that they are able to maintain normal or near-normal body temperature, respiratory function, and cardiovascular perfusion; that they are not subject to trauma inflicted by other animals; and that any lesions (e.g., wounds or masses) do not induce the animal to self-traumatize or otherwise needlessly subject the animal to significant pain or distress resulting from infection and/or inflammation.

Several common research uses of mice can be used to illustrate these principles:

1. **Cancer research**: Mice are frequently used to study a variety of cancer types. These studies can use mouse models in which autochthonous cancer develops as a result of experimental manipulation or as a consequence of genotype, while others use transplantation of cells, often from human tumors, into immunodeficient mice (Workman et al. 2010). In either case, tumors may arise that can grow to substantial proportions and infiltrate tissues with subsequent loss of tissue function. Large tumors, particularly those growing in the subcutaneous space, may interfere with the ability of the mouse to move, leading to an inability to access food and water. Fast-growing tumors may become necrotic and incite severe inflammation. Further, such tumors may expand and destroy neighboring tissues. For example, fast-growing subcutaneous tumors may press against the overlying skin, causing death and necrosis of the skin and resulting in an open tumor lesion. For all these reasons, it is important to have predetermined defined endpoints at which tumor-bearing mice are euthanized. Typical criteria may be based upon maximal tumor size (e.g., tumor diameter or volume as measured with a caliper); inability of the mouse to ambulate and access food and water; presence of necrotic or abraded, open skin overlying subcutaneous tumors; and unkempt appearance of the fur. Maximal tumor size is somewhat arbitrary, though limits of 20 mm in diameter or 2 cm^3 in volume are frequently applied. Although weight loss is sometimes used in developing endpoint criteria, and tumor-bearing mice may lack appetite and lose weight, the weight of some mice may actually be relatively normal

because of the added weight of the tumor mass; thus, the utility of body weight should be carefully considered when establishing clinical endpoint criteria for mice used in cancer research. Scoring of body condition and behavior is a useful approach, particularly for studies involving tumors that cannot be easily measured (Paster et al. 2009).

2. **Research with aged mice**: Research in which aged mice are used can be complicated by spontaneous disease associated with advancing age. For this reason, it is important to establish clinical endpoints that are acceptable from both ethical and scientific perspectives. General endpoints, such as those described previously, are a useful starting point, as are criteria for tumors as stated previously. A review of approaches to monitoring health and predicting death in elderly research mice is provided in Toth, 2018 (Toth 2018).

3. **Shock research**: Studies in which shock is experimentally induced may impose significant distress upon the animal. One important hallmark of shock is hypothermia; thus, measurement of body temperature and a predetermined temperature at which the animal is euthanized are useful components of a plan for setting clinical endpoints in research in which shock may result (Nemzek et al. 2004). Other criteria might include inability to ambulate, unkempt appearance, and body weight and overall body condition.

4. **Degenerative disease**: Mice are used as models for many degenerative diseases, including experimental autoimmune encephalopathy (a model for demyelinating disease such as multiple sclerosis), muscular dystrophy, and amyotrophic lateral sclerosis (ALS). As disease progresses, so do clinical signs, which can manifest most obviously as changes in mobility. Scoring systems based on a combination of appearance and behavior have been published for many of these models, which can aid in identifying humane endpoints (Guyenet al. 2010; Hatzipetros et al. 2015; Kobayashi et al. 2012; Miller and Karpus 2007).

Regardless of the specific criteria chosen, all members of the research team should understand the need to adhere to these criteria. Further, a specific chain of communication should be established so that, if animals need to be euthanized, appropriate samples might still be obtained. Further, all participants should understand in advance who has the final authority regarding the decision to euthanize an animal. Typically, such authority rests with the attending veterinarian and other members of the veterinary staff.

treating disease on a colony basis

Contagious diseases typically are viewed as colony problems rather than clinical problems for which individual mice are treated. When such diseases are diagnosed, the goals are to: (1) prevent further spread of the infection within the room or facility; (2) eliminate the infectious agent; and (3) preserve research, valuable experimental animals, and genetic resources to the greatest extent possible during pursuit of the first two goals.

Prevention of Spread

While the ultimate goal is to eliminate the unwanted microbial contaminant, this may take some time. Thus, the first efforts should be put toward preventing it from spreading to as-yet-uncontaminated areas of the room or facility. Success will depend on: (1) determining how far the infection was able to spread prior to detection; and (2) understanding, and effectively blocking, all ways in which the organism may spread.

To determine how far the infection has already spread, one must first think about all the possible connections between the infected colony and other colonies and areas. Records of animal movement are invaluable for assessing possible patterns of spread. For example, have animals from the infected colony been brought to a shared laboratory, procedure room, or testing area? Have any animals from the infected colony been moved to another animal room? Do any of the people who work with the infected colony also work with other colonies? Are there any supplies or pieces of equipment that were taken from the contaminated room and used in another room without being sterilized in between? If the answer to any of these questions is "yes," the potential exists that the infection has spread to these other areas. In this case, additional testing of animals in these areas is recommended.

Microbial contaminants and parasites vary quite a bit in terms of the ease and means by which they spread, with some spreading rapidly by different routes and others spreading quite slowly. While the measures taken can be significantly less rigorous with organisms that spread slowly, unless you have knowledge to the contrary, it is best to assume that a contaminant could spread swiftly and take actions to block all means by which this could happen. This means implementing measures to prevent spread by animals, people, and materials.

Animals: All microbial organisms and parasites can be spread by contaminated animals. It is best to stop all movement of live animals out of the contaminated room(s) and, also, to keep in mind that their tissues may be contaminated. Dead animals should be placed in a bag or container that will be treated with a disinfectant prior to being taken to an on-site incinerator or holding area. Removal should be by a predetermined route that avoids other animal housing or procedure areas to the extent possible. If the outbreak occurred within a room where microisolator or individually ventilated cages are in use, further spread within the room can be minimized by assuring that all personnel are using good *microisolator technique* (see section titled "Housing Systems" in Chapter 2).

People: People working with infected animals, or even in an area where they are being housed or studied, can have their clothing, hair, or even skin contaminated with the infectious agent. Especially in the case of hardy organisms that can live a long time in the environment, a person can serve as a fomite, or vehicle for transferring the contaminant from one mouse to another. To prevent this, people should wear protective clothing when working with potentially contaminated animals and should shed this clothing before leaving the room. It may be desirable to have people shower and change clothing after entering the affected room or area. Even if people shower and change clothes, it is best if they do not go into another animal housing or procedure area after leaving the contaminated room. Clothing, especially outer clothing and shoes, worn in the contaminated room should be treated as contaminated. Make sure that *all* people who have work assignments in the contaminated room—including investigative and custodial staff—are trained in these guidelines. Traffic patterns should be adjusted so that people do not enter any other animal rooms after entering the contaminated room, or, if they need to visit other rooms, they change clothes and possibly shower before doing so.

Materials: As with people, materials can serve as fomites and transfer contaminants from one animal or area to another. To the extent possible, it is best to minimize any movement of equipment or supplies out of a contaminated room. If something must be moved out, it should be treated with an appropriate disinfectant before leaving and should be discarded (in a safe location) or sterilized prior to use in another area or processed through the cage wash (many organisms will not survive the high temperatures in a cage wash, but some may). This caveat is particularly important in the case of caging and bedding from the contaminated room.

Approaches to Elimination

There are several possible approaches to eliminating infectious organisms. The best choice depends on the biology of the organism itself, as well as the circumstances in place in the facility. The choices include depopulation, test and cull, burnout, and treatment with medications.

Depopulation

While generally not the most appealing option, *depopulation*—elimination of all animals in the contaminated room—is the most definitive option. This is particularly true when the infection is widespread within a room or colony and/or the organism spreads quickly or can persist for long periods of time in the environment. Depopulation can involve euthanizing all animals in the room or colony, but it can also involve simply moving them to another location where the contaminant can be tolerated. If the animals are relocated, it is important to take precautions to make sure that the contaminant is not brought back to the original location through uncontrolled movements of animals, people, or materials. Following depopulation, appropriate environmental decontamination steps must be taken to ensure the agent does not persist prior to re-housing animals in the space.

Test and cull

With this approach, animals are tested individually to determine whether they are infected, and positive animals (and their cagemates) are eliminated from the colony. Because the contaminant may continue to spread through the colony while testing is being completed, and because testing may fail to detect all currently infected animals, one round of test and cull may not be adequate to ensure elimination of the contaminant. Testing should continue until all animals sampled are negative, and two to three rounds of all-negative results improve confidence that the infectious agent has been eliminated. In general, it is advisable to wait 1–2 weeks between successive rounds of testing to allow for spread of the agent to manifest in shedding or seroconversion. With most infectious agents, introduction of new animals into the colony—including breeding—must cease for test and cull to work effectively.

Burnout

In the past, this was viewed as a feasible option for eradicating infectious viruses such as mouse hepatitis virus that spread rapidly, are cleared quickly by infected immunocompetent animals, and induce immunity to reinfection. The concept is to isolate the infected colony—no breeding or introduction of new animals—and allow the infectious agent to make its way through the entire population. Animals are tested for antibodies to the contaminant and are considered to be free of the virus within 4 weeks of testing antibody positive. Four weeks after all animals in the room test antibody positive, the room can be considered free of infectious virus. While this approach is still an option if all of the previously mentioned conditions pertain, the likelihood that all these conditions do pertain is greatly reduced in modern mouse colonies as compared with those of the past. The problem is the prevalence of immune-deficient and immunologically abnormal mice in modern colonies. The immune status of many genetically engineered mice is not fully known. Viruses that are rapidly cleared by immunologically normal animals may not be cleared quickly, or at all, by animals with immunological abnormalities. Such animals may remain infected, and capable of shedding infectious virus, even after becoming antibody positive. Therefore, burnout should be approached with caution and considered as an option only when all animals in the colony are *known* to be *fully* immune competent.

Medical treatment

There are no medications available for the treatment and eradication of viral pathogens, but many parasites and some bacteria and fungi can be controlled, and even eliminated, by drug treatments. In general, for drugs to be useful, they must be given to each animal by an appropriate route and at an appropriate dose. They also must be given for an appropriate length of time and/or repeated. In the case of some organisms, pinworms for example, it may also be necessary to coordinate treatment of the animals with careful disinfection of the environment. Whenever medications are used, it is important to verify after treatment is concluded—ideally, several weeks or months after treatment is concluded—that the infectious organism has been truly eliminated and not just temporarily reduced to undetectable levels.

Facility Decontamination

As mentioned above, thorough decontamination of the environment is an important step following depopulation and may be an essential aspect of disease eradication by other means. It is most critical when dealing with an infectious agent that is able to remain viable in the environment for prolonged periods of time. Before reintroducing clean animals into a room that has been depopulated, all supplies and nonfixed equipment and furnishings should be removed. Unless such items can be autoclaved or otherwise sterilized, they should be treated as contaminated waste or redeployed in an area where the infectious agent could be tolerated. The room, including any fixed equipment, should be manually cleaned and sanitized. Follow-up treatment with a high-level disinfectant or chemical sterilant is essential to assure elimination of environmentally persistent contaminants. Careful cleaning and disinfection of the environment are also recommended if animals will remain in the room following a test-and-cull effort, but any chemical used should be one that will pose no health threat to the mice.

references

Adams SC, Felt SA, Geronimo JT, Chu DK. 2016. A "Pedi" cures all: toenail trimming and the treatment of ulcerative dermatitis in mice. *PloS One* **11**(1):e0144871.

Adams SC, Myles MH, Tracey LN, Livingston RS, Schultz CL, Reuter JD, Leblanc M. 2019. Effects of pelleting, irradiation, and autoclaving of rodent feed on MPV and MNV infectivity. *J Am Assoc Lab Anim Sci* **58**:542–550.

Adhikary S, Nicklas W, Bisgaard M, Boot R, Kuhnert P, Waberschek T, Aalbaek B, Korczak B, Christensen H. 2017. *Rodentibacter* gen. nov. including *Rodentibacter pneumotropicus* comb. nov., *Rodentibacter heylii* sp. nov., *Rodentibacter myodis* sp. nov., *Rodentibacter ratti* sp. nov., *Rodentibacter heidelbergensis* sp. nov., *Rodentibacter trehalosifermentans* sp. nov., *Rodentibacter rarus* sp. nov., *Rodentibacter mrazii* and two genomospecies. *Int J Syst Evol Microbiol* **67**:1793–1806.

Aherrahrou Z, Doehring LC, Ehlers EM, Liptau H, Depping R, Linsel-Nitschke P, Kaczmarek PM, Erdmann J, Schunkert H. 2008. An alternative splice variant in *Abcc6*, the gene causing dystrophic calcification, leads to protein deficiency in C3H/He mice. *J Biol Chem* **283**:7608–7615.

Altholtz LY, La Perle KM, Quimby FW. 2006. Dose-dependent hypothyroidism in mice induced by commercial trimethoprim-sulfamethoxazole rodent feed. *Comp Med* **56**:395–401.

Alvarado CG, Franklin CL, Dixon LW. 2016. Retrospective evaluation of nail trimming as a conservative treatment for ulcerative dermatitis in laboratory mice. *J Am Assoc Lab Anim Sci* **55**:462–466.

Becker SD, Bennett M, Stewart JP, Hurst JL. 2007. Serological survey of virus infection among wild house mice (*Mus domesticus*) in the UK. *Lab Anim* **41**:229–238.

Bracken TC, Cooper CA, Ali Z, Truong H, Moore JM. 2017. *Helicobacter* infection significantly alters pregnancy success in laboratory mice. *J Am Assoc Lab Anim Sci* **56**:322–329.

Bremell T, Lange S, Svensson L, Jennische E, Gröndahl K, Carlsten H, Tarkowski A. 1990. Outbreak of spontaneous staphylococcal arthritis and osteitis in mice. *Arthrit Rheumat* **33**:1739–1744.

Burkholder T, Foltz C, Karlsson E, Linton CG, Smith JM. 2012. Health evaluation of experimental laboratory mice. *Curr Protocols Mouse Biol* **2**:145–165.

Cates CC, McCabe JG, Lawson GW, Couto MA. 2014. Core body temperature as adjunct to endpoint determination in murine median lethal dose testing of rattlesnake venom. *Comp Med* **64**:440–447.

Charles River Laboratories. 2020. Rodent & rabbit infectious agents. Secondary rodent & rabbit infectious agents. [Cited 10.19.2020

2020]. Available at: https://www.criver.com/products-services /research-models-services/animal-health-surveillance/infectious-agent-information?region=3601.

Clark SE, Purcell JE, Bi X, Fortman JD. 2017. Cross-foster rederivation compared with antibiotic administration in the drinking water to eradicate *Bordetella pseudohinzii*. *J Am Assoc Lab Anim Sci* **56**:47–51.

Clark SE, Purcell JE, Sammani S, Steffen EK, Crim MJ, Livingston RS, Besch-Williford C, Fortman JD. 2016. *Bordetella pseudohinzii* as a confounding organism in murine models of pulmonary disease. *Comp Med* **66**:361–366.

Compton SR, Booth CJ, Macy JD. 2017. Murine astrovirus infection and transmission in neonatal CD1 mice. *J Am Assoc Lab Anim Sci* **56**:402–411.

Dammann P, Hilken G, Hueber B, Kohl W, Bappert MT, Mahler M. 2011. Infectious microorganisms in mice (*Mus musculus*) purchased from commercial pet shops in Germany. *Lab Anim* **45**:271–275.

Dyson MC, Eaton KA, Chang C. 2009. *Helicobacter* spp. in wild mice (*Peromyscus leucopus*) found in laboratory animal facilities. *J Am Assoc Lab Anim Sci* **48**:754–756.

Easterbrook JD, Kaplan JB, Glass GE, Watson J, Klein SL. 2008. A survey of rodent-borne pathogens carried by wild-caught Norway rats: a potential threat to laboratory rodent colonies. *Lab Anim* **42**:92–98.

Fiebig K, Jourdan T, Kock MH, Merle R, Thöne-Reineke C. 2018. Evaluation of infrared thermography for temperature measurement in adult male NMRI nude mice. *J Am Assoc Lab Anim Sci* **57**:715–724.

Glass AM, Coombs W, Taffet SM. 2013. Spontaneous cardiac calcinosis in BALB/cByJ mice. *Comp Med* **63**:29–37.

Goelz MF, Thigpen JE, Mahler J, Rogers WP, Locklear J, Weigler BJ, Forsythe DB. 1996. Efficacy of various therapeutic regimens in eliminating *Pasteurella pneumotropica* from the mouse. *Lab Anim Sci* **46**:280–285.

Gopinath D. 2018. Splendore-Hoeppli phenomenon. *J Oral Maxillofac Pathol* **22**:161–162.

Guyenet SJ, Furrer SA, Damian VM, Baughan TD, La Spada AR, Garden GA. 2010. A simple composite phenotype scoring system for evaluating mouse models of cerebellar ataxia. *J Visual Exp* **39**:1787.

Hampton AL, Aslam MN, Naik MK, Bergin IL, Allen RM, Craig RA, Kunkel SL, Veerapaneni I, Paruchuri T, Patterson KA, Rothman ED, Hish GA, Varani J, Rush HG. 2015. Ulcerative dermatitis in C57BL/6NCrl mice on a low-fat or high-fat diet with or without a mineralized red-algae supplement. *J Am Assoc Lab Anim Sci* **54**:487–496.

Hatzipetros T, Kidd JD, Moreno AJ, Thompson K, Gill A, Vieira FG. 2015. A quick phenotypic neurological scoring system for evaluating disease progression in the SOD1-G93A mouse model of ALS. *J Visual Exp* **104**:53257.

Held N, Hedrich HJ, Bleich A. 2011. Successful sanitation of an EDIM-infected mouse colony by breeding cessation. *Lab Anim* **45**:276–279.

Ike F, Sakamoto M, Ohkuma M, Kajita A, Matsushita S, Kokubo T. 2016. *Filobacterium rodentium* gen. nov., sp. nov., a member of *Filobacteriaceae* fam. nov. within the phylum *Bacteroidetes*; includes a microaerobic filamentous bacterium isolated from specimens from diseased rodent respiratory tracts. *Int J System Evolution Microbiol* **66**:150–157.

Kjeldsberg E, Hem A. 1985. Detection of astroviruses in gut contents of nude and normal mice. Brief report. *Arch Virol* **84**:135–140.

Kobayashi YM, Rader EP, Crawford RW, Campbell KP. 2012. Endpoint measures in the mdx mouse relevant for muscular dystrophy pre-clinical studies. *Neuromusc Disorders* **22**:34–42.

Kort WJ, Hekking-Weijma JM, TenKate MT, Sorm V, VanStrik R. 1998. A microchip implant system as a method to determine body temperature of terminally ill rats and mice. *Lab Anim* **32**:260–269.

Langford DJ, Bailey AL, Chanda ML, Clarke SE, Drummond TE, Echols S, Glick S, Ingrao J, Klassen-Ross T, Lacroix-Fralish ML, Matsumiya L, Sorge RE, Sotocinal SG, Tabaka JM, Wong D, van den Maagdenberg AM, Ferrari MD, Craig KD, Mogil JS. 2010. Coding of facial expressions of pain in the laboratory mouse. *Nat Methods* **7**:447–449.

Marx JO, Gaertner DJ, Smith AL. 2017. Results of survey regarding prevalence of adventitial infections in mice and rats at biomedical research facilities. *J Am Assoc Lab Anim Sci* **56**:527–533.

Marx JO, Vudathala D, Murphy L, Rankin S, Hankenson FC. 2014. Antibiotic administration in the drinking water of mice. *J Am Assoc Lab Anim Sci* **53**:301–306.

Matsumiya LC, Lavoie C. 2003. An outbreak of *Pasteurella pneumotropica* in genetically modified mice: treatment and elimination. *Contemp Top Lab Anim Sci* **42**:26–28.

Michaud CR, Qin J, Elkins WR, Gozalo AS. 2016. Comparison of 3 topical treatments against ulcerative dermatitis in mice with a C57BL/6 background. *Comp Med* **66**:100–104.

Miller SD, Karpus WJ. 2007. Experimental autoimmune encephalomyelitis in the mouse. *Curr Protocols Immunol* **77**:15.11.11–15.11.18.

Nashat MA, Luchins KR, Lepherd ML, Riedel ER, Izdebska JN, Lipman NS. 2017. Characterization of *Demodex musculi* infestation, associated comorbidities, and topographic distribution in a mouse strain with defective adaptive immunity. *Comp Med* **67**:315–329.

Nashat MA, Ricart Arbona RJ, Lepherd ML, Santagostino SF, Livingston RS, Riedel ER, Lipman NS. 2018a. Ivermectin-compounded feed compared with topical moxidectin-imidacloprid for eradication of *Demodex musculi* in laboratory mice. *J Am Assoc Lab Anim Sci* **57**:483–497.

Nashat MA, Ricart Arbona RJ, Riedel ER, Francino O, Ferrer L, Luchins KR, Lipman NS. 2018b. Comparison of diagnostic methods and sampling sites for the detection of *Demodex musculi*. *J Am Assoc Lab Anim Sci* **57**:173–185.

Nemzek JA, Xiao HY, Minard AE, Bolgos GL, Remick DG. 2004. Humane endpoints in shock research. *Shock* **21**:17–25.

Ng TF, Kondov NO, Hayashimoto N, Uchida R, Cha Y, Beyer AI, Wong W, Pesavento PA, Suemizu H, Muench MO, Delwart E. 2013. Identification of an astrovirus commonly infecting laboratory mice in the US and Japan. *PLoS One* **8**:e66937.

Parker SE, Malone S, Bunte RM, Smith AL. 2009. Infectious diseases in wild mice (*Mus musculus*) collected on and around the University of Pennsylvania (Philadelphia) Campus. *Comp Med* **59**:424–430.

Paster EV, Villines KA, Hickman DL. 2009. Endpoints for mouse abdominal tumor models: refinement of current criteria. *Comp Med* **59**:234–241.

Pearson EC, Pugazhenthi U, Fong DL, Smith DE, Nicklawsky AG, Habenicht LM, Fink MK, Leszczynski JK, Schurr MJ, Manuel CA. 2020. Metaphylactic antibiotic treatment to prevent the transmission of *Corynebacterium bovis* to immunocompromised mouse offspring. *J Am Assoc Lab Anim Sci* **59**:712–718.

Pritchett-Corning KR, Cosentino J, Clifford CB. 2009. Contemporary prevalence of infectious agents in laboratory mice and rats. *Lab Anim* **43**:165–173.

Radaelli E, Arnold A, Papanikolaou A, Garcia-Fernandez RA, Mattiello S, Scanziani E, Cardiff RD. 2009. Mammary tumor phenotypes in wild-type aging female FVB/N mice with pituitary prolactinomas. *Vet Pathol* **46**:736–745.

Roble GS, Gillespie V, Lipman NS. 2012. Infectious disease survey of *Mus musculus* from pet stores in New York City. *J Am Assoc Lab Anim Sci* **51**:37–41.

Sellers RS, Clifford CB, Treuting PM, Brayton C. 2012. Immunological variation between inbred laboratory mouse strains: points to consider in phenotyping genetically immunomodified mice. *Vet Pathol* **49**:32–43.

Sheldon WG, Curtis M, Kodell RL, Weed L. 1983. Primary harderian gland neoplasms in mice. *J Nat Cancer Inst* **71**:61–68.

Shinohara Y, Frith CH. 1986. Correlation between spontaneous and experimentally induced tumors in female BALB/c mice in a large 2-acetylaminofluorene study. *J Environ Pathol Toxicol Oncol* **6**:85–95.

Smith PC, Zeiss CJ, Beck AP, Scholz JA. 2016. *Demodex musculi* infestation in genetically immunomodulated mice. *Comp Med* **66**:278–285.

Tappe-Theodor A, King T, Morgan MM. 2019. Pros and cons of clinically relevant methods to assess pain in rodents. *Neurosci Biobehav Rev* **100**:335–343.

Toth LA. 2018. Identifying and implementing endpoints for geriatric mice. *Comp Med* **68**:439–451.

Towne JW, Wagner AM, Griffin KJ, Buntzman AS, Frelinger JA, Besselsen DG. 2014. Elimination of *Pasteurella pneumotropica* from a mouse barrier facility by using a modified enrofloxacin treatment regimen. *J Am Assoc Lab Anim Sci* **53**:517–522.

Ullman-Cullere MH, Foltz CJ. 1999. Body condition scoring: a rapid and accurate method for assessing health status in mice. *Lab Anim Sci* **49**:319–323.

University of Missouri - Comparative Medicine Program and IDEXX-BioAnalytics. 2020. DISEASES OF RESEARCH ANIMALS - DORA. Secondary DISEASES OF RESEARCH ANIMALS - DORA. [Cited 10.19.2020 2020]. Available at: http://dora.missouri.edu/mouse/.

Ward JM. 2006. Lymphomas and leukemias in mice. *Exp Toxicol Pathol* **57**:377–381.

Whiteside TE, Qu W, DeVito MJ, Brar SS, Bradham KD, Nelson CM, Travlos GS, Kissling GE, Kurtz DM. 2020. Elevated arsenic and lead concentrations in natural healing clay applied topically as a treatment for ulcerative dermatitis in mice. *J Am Assoc Lab Anim Sci* **59**:212–220.

Won YS, Jeong ES, Park HJ, Lee CH, Nam KH, Kim HC, Hyun BH, Lee SK, Choi YK. 2006. Microbiological contamination of laboratory mice and rats in Korea from 1999 to 2003. *Exp Anim* **55**:11–16.

Workman P, Aboagye EO, Balkwill F, Balmain A, Bruder G, Chaplin DJ, Double JA, Everitt J, Farningham DA, Glennie MJ, Kelland LR, Robinson V, Stratford IJ, Tozer GM, Watson S, Wedge SR, Eccles SA. 2010. Guidelines for the welfare and use of animals in cancer research. *Br J Cancer* **102**:1555–1577.

preventive medicine

receiving

The process for receiving mice from another institution, or from another area within the same institution, is among the most critical aspects of a program for preventive medicine. If designed and implemented with care, the receiving process can do much to prevent infected animals or pathogens from being transferred into the animal barrier.

Reviewing Health Reports

A health report is a documented summary of the data obtained from health surveillance of the colony of origin. It is not a laboratory testing report, which is the raw results provided by the testing laboratory. While it is no guarantee of the health status of the mice received from another institution, it can provide a good indication of the facility's health monitoring practices and, in many cases, the health history of the colony of origin. Ideally, a health report should be obtained and reviewed *before* the animals are shipped; this can help you determine in advance whether the animals are likely to meet the health standards of your facility. If the health report shows that the colony of origin is infected with organisms on your exclusion list, or if they do not routinely test for organisms on your exclusion list, you typically have the opportunity to cancel the shipment, request additional testing at the institution of origin prior to shipment, or develop a plan to quarantine and/or rederive mice when they arrive. If additional health information is shipped with the mice—for example, a more

DOI: 10.1201/9780429353086-5

recent health report—it is prudent to review this new information as soon as the animals arrive, especially if the plan was to send them directly to the barrier. When reviewing a health report, look for the following information.

1. An indication of what area(s) of the facility is/are covered by the report, i.e., does it show health monitoring data from a specific room or area within the facility or the entire facility? If it shows data from a specific room or area, verify that this is the room/area of origin of the mice that will be/have been shipped to your facility.

2. Health monitoring information for all infectious agents that you wish to exclude from the barrier. It is common to monitor the colonies for some organisms more frequently than others (see section titled "Health Surveillance and Monitoring"), but absence of any information about an organism of concern could be a red flag. At the least, it should provoke a call to the veterinarian at the institution of origin.

3. A health monitoring history that extends back at least 6 months, preferably longer. Repeated negative results for an infectious agent increase confidence that the colony is truly negative for that agent.

4. An indication of the number of animals tested. Confidence that the colony is negative for a particular organism is greater if several mice were tested than if only one or two mice were tested.

5. An indication of which types of animals or materials were tested (e.g., colony animals, dirty bedding sentinels, contact sentinels, exhaust air dust). Because different agents have different routes of transmission and transmissibility, a consideration of testing source is important to determine if a negative result truly reflects the absence of disease in the colony, rather than a failure to detect the agent.

6. An indication of what tissue was tested for each infectious agent and the type of test used. Consult your facility veterinarian to verify the appropriateness of the tissues tested and methodology.

7. The testing laboratory: If you do not recognize the laboratory, or if the testing was done in-house, it is worth some additional effort to verify that the laboratory has expertise in

testing for rodent pathogens and that it follows appropriate quality control measures.

There is no uniformity in laboratory or health report design among different institutions and testing laboratories, and so it can be challenging to find all of the pertinent information when reviewing an unfamiliar report format. Read carefully through the entire report, as important information may not stand out. If information seems to be missing, or if it is not readily interpretable, contact the veterinarian at the institution of origin for clarification.

Initial handling of animals at the final destination depends in part on their origin. For animals being received from another institution, the following procedure is recommended.

1. Examine the shipping container for damage or evidence that the filter material has become excessively wet. Either condition suggests that the barrier within the container has been breached and that the animals may have been exposed to pathogens in the outside environment.

2. As the outside of the shipping container may have been exposed to unwanted microbial agents, it is desirable to wipe down the box with a high-level disinfectant prior to opening. This step is particularly important if the animals will be transferred directly into a barrier environment.

3. During unpacking, inspect each animal to make sure that: (1) it is the strain, sex, and age/weight expected; and (2) that it appears healthy (consistent with the expected phenotype). Keep in mind, however, that mice typically will lose some weight during shipping, particularly when shipped over longer distances and/or by air. If the animals are not as expected, or if they appear ill or injured, it may be appropriate to contact the vendor or institution of origin and notify it of your concerns.

Mice that were transported from another area within your own institution should not be allowed to sit for long periods without food and water. If they are to be necropsied or used in a study, they should be utilized promptly or, if this is not possible, they should be given food and water and placed in a secure, well-ventilated, climate-controlled holding area. For receipt of mice into a new barrier, careful attention must be paid to disinfecting the cage or transport container at the barrier entrance. Alternatively, for intra-institutional transfer, the cage(s) can be securely enclosed within an impenetrable outer container, which can then be disinfected and/or removed at the barrier entrance. Care must be taken to ensure that animals are kept in the impenetrable container for the minimum possible time to prevent issues related to lack of ventilation.

Once animals have been unpacked and placed into new, clean cages, they can be transported directly to the barrier, they can be quarantined, or they can be rederived. The decision of which of these options to use is based on weighing the perceived risk and tolerance for risk against practical considerations such as cost and space. Many institutions use two, or even all three, of these options depending on the source of the mice. So, for example, mice from a trusted source accompanied by a clean health report may be sent directly to the barrier. Mice from other sources may be quarantined, or even rederived, depending on perceived risk and risk tolerance.

Options for Newly Imported Mice

Transport directly to barrier

This is the simplest, least expensive option, and it allows investigators immediate access to their mice. However, it is the riskiest option, as there is no verification of the animals' health status prior to introduction into the established colony. This option is used most often for mice received from "approved" vendors that utilize strict barrier practices at their facilities and periodically provide colony health information. (*Note*: Because of the effects of shipping stress on mouse physiology, it is strongly recommended that mice not be used for most experimental purposes for at least 3–5 days after receipt.)

Quarantine

This option requires space, resources, and time for the quarantine program. Depending on how it is set up, inability of investigators to access their mice for a significant period of time could have significant impacts on research. However, the risk of introducing diseased animals into the colony is reduced.

Principles

Quarantine programs are most effective when the quarantine area is located in a different part of the facility from established mouse colonies. While different shipments may be housed within the same quarantine facility, it is preferable to separate them from each other in some manner, e.g., place them on different racks or in different isolators, cubicles, bio-bubbles, etc. If more than one shipment will be quarantined in the same secondary enclosure (e.g., room or cubicle), mice should be kept in microisolator or ventilated cages.

To prevent transmission of pathogens from quarantined animals to established colonies, caretakers must never work with established mouse colonies after handling rodents in quarantine. This may be avoided by assigning work in the quarantine area to a dedicated caretaker or to a caretaker who works primarily with non-rodent species. If neither of these options is feasible, care of quarantined animals should be done at the end of the day.

The length of quarantine represents a balance between the desire to avoid introducing any new pathogens into an established colony and the desire to minimize the negative impact on research of a lengthy quarantine. Animals can be tested immediately upon arrival, but this will not identify infections that occurred shortly before or during shipment. The timing between arrival and testing depends on several factors, including the type of testing conducted (PCR vs. serology) and whether testing will be performed on colony animals or sentinels. The absolute minimum time needed to reliably detect such recent infections is 3 weeks, and 4–6 weeks is better, if dirty bedding sentinels will be used for testing (see section titled "Health Surveillance and Monitoring").

Practices

While many of the infectious organisms that are unwanted in modern mouse colonies do not cause overt disease, it is important that caretakers be particularly vigilant in observing quarantined animals for signs of illness. If any such signs are seen, contact the facility veterinarian immediately. Even in the absence of clinical signs or a known history of infestation, many institutions have a policy of treating animals in quarantine for internal and/or external parasites. At the direction of the facility veterinarian, treatment with a pesticide such as fenbendazole or ivermectin might be initiated.

Mice should not be released from quarantine until representative animals have tested negative for all organisms on the facility's exclusion list. Animals submitted for testing may be colony animals, contact sentinels, or dirty bedding sentinels. If dirty bedding sentinel mice will be utilized for this purpose, it is important to make sure that the sentinels receive adequate exposure to potentially infectious material in the used bedding (see section titled "Selection of Test Subjects").

Acute use quarantine

A disadvantage of a well-designed quarantine program is that investigators must wait several weeks to get their mice. This may be more than a simple inconvenience if the experimental protocol requires use of the mice soon after arrival, e.g., it requires use of timed pregnant animals. An "acute use" quarantine program can be set up to allow investigators to manipulate mice in quarantine. There are a number of ways in which this can be done, but, to be effective, the program must maintain the separation between animals in quarantine and established colonies. One approach that works well is to provide dedicated procedure space immediately adjacent to the quarantine area. Quarantine mice may be taken to this procedure space but nowhere else. Ideally, investigators who use this space should not enter established colony rooms or any area where mice from established colonies are handled prior to returning to the animal room. At the very least, they should not enter such areas directly after working with mice in quarantine.

Rederivation

With a well-designed and implemented rederivation program, the risk of introducing infectious agents into the colony is reduced to the lowest extent possible. It is more expensive than quarantine and it requires additional space and technical expertise. Depending on the method of rederivation used, the delay between initial receipt of mice and release of rederived offspring may be several months or longer. The most commonly used methods of rederivation are newborn fostering, hysterectomy derivation, and embryo transfer. All of these methods require a source of foster mothers that are confirmed free of all pathogens on the facility's exclusion list. Depending on the frequency with which rederivations are performed, it may be worth keeping a colony that can provide a steady supply of foster mothers. If rederivations will be performed infrequently, foster mothers can be selected and prepared or ordered prior to each procedure. For newborn fostering or hysterectomy derivation, timed matings are planned such that the foster mother delivers her litter 1–3 days prior to the anticipated rederivation. For embryo transfer, adult females are rendered pseudopregnant as described later in this chapter.

Newborn fostering

Lactating mice will often accept pups from another litter and raise them as their own. When the litter is transferred to the foster mother immediately after birth, this approach can be useful for eliminating pathogens and parasites that are transferred from adult to offspring after birth. It is not a reliable method for eliminating pathogens that may be transferred before or during birth, and is not as reliable as hysterectomy derivation or embryo transfer for eliminating many other pathogens. Its advantage over these other techniques is that it requires little technical expertise.

Procedure: The foster mother should be an experienced mother from a strain that is known for good maternal behavior and, preferably, for having larger litters. Her litter should be within 1-2d of age of the litter to be fostered. Remove her from her litter and place her in a new cage. Then, euthanize her litter and mingle the litter to be fostered in the dirty bedding from her old cage. Put the mother back in the cage with the foster pups. Some pups from her own litter may be kept with the foster pups *provided that*: (1) the total number will not exceed the original litter size by more than one or two, and (2) the foster pups will have a different coat color than the original pups. Check back on the litter an hour afterward to make sure that the mother has accepted the foster pups, as indicated by the observation that she has gathered them together into the nest.

Hysterectomy derivation

This method involves removal of the pregnant uterus from an infected female followed by transfer of the revived pups to an uninfected foster mother (Lipman et al. 1987; Pilgrim and Parks 1968). When performed correctly, hysterectomy derivation can be used successfully to eliminate a wide range of pathogens.

Procedure: As close as possible to the time of expected natural parturition, euthanize the pregnant mouse whose pups are to be rederived. Euthanasia is best performed by physical means, such as cervical dislocation, since drugs and volatile anesthetics may influence the viability of the pups, particularly those of fragile strains. The uterus is removed using aseptic procedures and transferred to a sterile surface. The pups are then carefully removed from the incised uterus, usually within a biological safety cabinet, a laminar flow hood, or an isolator. Excess mucus and secretions can be removed from the face of the pups with a sterile, moistened swab, and respiration can be stimulated by gentle rubbing of the pups with soft, sterile gauze. Once the pups are bright pink-to-red and squeaking, follow the procedure described previously for transfer to the foster mother.

Embryo transfer

When performed with skill and care, rederivation through embryo transfer has been proven effective in eliminating as wide a range of pathogens as hysterectomy derivation. Note, however, that this approach does not work equally well for mice of all strains and genetic backgrounds. The specific procedures are described in detail elsewhere (Behringer et al. 2014), but can involve either collection of embryos from donor mice (Reetz et al. 1988) or superovulation followed by in vitro fertilization (Suzuki et al. 1996). In either case, embryos are transferred to specific pathogen–free (SPF) pseudopregnant females under aseptic conditions. Pseudopregnant mice are produced by mating young (at least 6 weeks) adult females with vasectomized males. If possible, verify that the vasectomized males are sterile before mating. If this cannot be done, it is recommended that both the pseudopregnant female and vasectomized male be selected with a thought to assuring that any pups that might result from a fertile mating would be of a different coat color than the pups that will result from the embryo transfer.

Cryopreservation

Cryopreservation can be used to preserve, and assure a clean supply of, valuable genetic stocks of mice. Techniques are available to cryopreserve—and recover—embryos at various stages of preimplantation development, as well as sperm, spermatogonial stem cells, ova, and ovaries. Detailed discussions of these techniques, with specific instructions for performing them, are readily available (Behringer et al. 2014). The efficiency of recovery of these materials from cryopreservation varies with both the strain of mouse and the material frozen. Consult an expert before deciding which material(s) and technique(s) will be used to cryopreserve a particular strain.

testing of biological materials

Risks for Humans and Animals

Biological materials are widely used in mice in biomedical research. These include transplantable tumors, body fluids, cell lines, hybridomas, virus stocks, antibody preparations, and tissue scaffolding. Any material of biological origin, i.e., any tissue or other material obtained from an animal or human, may be contaminated with unwanted microbial agents. Such materials may be contaminated with agents that infected the animals (or humans) from which they came initially, or may become contaminated by subsequent passage through infected animals. In practice, materials that are most likely to be contaminated with rodent pathogens are those that originated in, or were passaged through, rodents. Human viruses are most likely to contaminate materials taken from humans, especially fresh xenografts (tissues transplanted directly from human patients to mice without prior establishment of in vitro cell culture or tissue culture). Microbial agents in biological materials are potentially hazardous to inoculated or engrafted animals, as well as to their human handlers. Rodent pathogens that may contaminate these materials are summarized in Table 5.1, and possible human viral contaminants are listed in Table 5.2. Some of these agents have caused significant disease in rodents as a result of inoculation of contaminated materials. Some of the agents are zoonotic (can infect rodents and humans) and pose a risk to human handlers, especially when contaminated materials are inoculated into immunodeficient animals that develop productive infections and shed the infective agents copiously. Many of these contaminants can be confounding variables on research results, even in the absence of clinical disease. Quality assurance and quality control (QA/QC) information for commercially obtained materials should be assessed carefully to determine what testing has been performed. Further testing often is warranted to protect personnel as well as valuable animals and experiments. Typically, investigators are charged with ensuring that appropriate

TABLE 5.1: RODENT PATHOGENS THAT MAY CONTAMINATE BIOLOGICAL MATERIALS

Ectromelia virus	Mouse thymic virus (MTLV)
Lactic dehydrogenase elevating virus (LDEV)	Murine norovirus (MNV)
Lymphocytic choriomeningitis virus (LCMV)	*Mycoplasma* spp.
Mouse adenovirus (MAV1,2)	Parvoviruses (MMV, MPV, or MKPV/ MuCPV)
Mouse cytomegalovirus (MCMV)	Pneumonia virus of mice (PVM)
Mouse hepatitis virus (MHV)	Reovirus (Reo1,3)
Mouse polyoma virus (MPyV)	Sendai virus
Mouse pneumotropic virus (K virus)	Theiler's mouse encephalomyelitis virus (TMEV)
Mouse rotavirus (EDIM)	

TABLE 5.2: HUMAN VIRUSES THAT MAY CONTAMINATE BIOLOGICAL MATERIALS

Adeno-associated virus	Human herpesvirus 8 (KSHV Kaposi's sarcoma-associated herpesvirus)
Hantavirus	Human papillomavirus types 16 or 18 (a cause of human cervical cancer)
Hepatitis viruses A, B, C	Human papillomavirus 18
Hepatitis B virus	Human parvovirus B19
Human foamy virus (retrovirus)	Human immunodeficiency virus (HIV)
Human herpesvirus 4 (EBV, Epstein Barr virus)	Human polyomavirus JC, BK
Human herpesvirus 5 (human CMV)	Human T lymphotropic virus (retrovirus cause of some adult T cell leukemias/lymphoma)
Human herpesvirus 6 (herpes lymphotropic virus)	Lymphocytic choriomeningitis virus (LCMV)

testing is performed on biologics used in research animals. Testing requirements should apply to any materials that originate in rodents or have been exposed to rodents directly (via in vivo passage) or indirectly (via tissue culture media additives). Exemptions may exist for materials that originated in or have been passaged through animals on campus housed at the same or higher health status than the animals to which the materials will be administered, as presumably any contamination would have been identified from routine health surveillance of that colony. Many commercial diagnostic labs provide PCR panels specifically for testing biological materials and should be consulted for tests offered and the best samples to submit.

health surveillance and monitoring

Periodic evaluation and documentation of the microbial status of the colony are critical components of the preventive medicine program. The specific approach used will vary with the culture and goals of the institution.

Principles

The purpose of health surveillance is to provide evidence that the colony is of the expected microbial status or, conversely, to find at least one infected animal if an organism from your exclusion list has gotten into the colony. Any surveillance program must balance cost and other practical considerations against the extent to which an as-yet-undetected microbial contamination can be tolerated. No health surveillance program can provide absolute certainty regarding the health status of the colony. Even if the program were to involve testing every animal in the colony, infectious agents could go undetected because of the uncertainties that are inherent in testing. Confidence in the health status of the colony is improved by increasing both the frequency of monitoring and the number of samples collected at each time point. In addition, a well-designed surveillance program will take into account selection of test subject and testing methods or selection of a testing laboratory.

Frequency of monitoring: Frequency of testing varies with the situation. In a small, closed facility with an excellent pest control program in which new animals are seldom introduced from outside, a single annual evaluation, along with evaluation of any animals that become ill, might be sufficient. In production facilities, where documentation of "clean" or SPF status is critical, short *core* panels of evaluation may be done every 2 weeks, with larger panels performed on a less frequent basis. Most research facilities fall somewhere in between these two extremes of testing frequency, and quarterly evaluations are common. Facilities with long exclusion lists often choose to test for organisms of greater concern more frequently than organisms of lesser concern. The organisms of greater concern may include those that are most commonly encountered in modern mouse colonies, e.g., mouse parvoviruses, as well as those that have the greatest potential to interfere with the research being conducted at the facility.

Selection of Test Subjects

Health monitoring may be performed using sentinel mice, colony mice, or environmental samples such as cage lids or exhaust air dust.

Sentinels are mice that have been introduced into the colony for the specific purpose of health monitoring; they include dirty bedding sentinels (those that have been exposed to soiled bedding from the colony) and contact sentinels, which are placed directly into the cages of colony animals. Sentinels are advantageous in that they can be chosen specifically for desirable characteristics such as age and immune status. On the negative side, transmission of pathogens by soiled bedding may be both slow and incomplete, thereby rendering dirty bedding sentinels less reliable indicators of the true health status of the colony (this is less of an issue with contact sentinels) (Compton et al. 2004; Dillehay et al. 1990; Manuel et al. 2008; Smith et al. 2007; Thigpen et al. 1989). This is especially true for pathogens that require close contact for transmission, such as respiratory viruses and fur mites. Also, there is always a chance that sentinels could actually introduce an unwanted contaminant into the colony. To enhance the likelihood of pathogen transmission to dirty bedding sentinels, place a good amount of soiled bedding from each colony cage into the sentinel cage at each cage change (Besselsen et al. 2008; Thigpen et al. 1989). Allow a minimum of 4–6 weeks of such exposure, with repeated transfers of dirty bedding, before testing the sentinels. A surveillance program based primarily on dirty bedding sentinels can be improved by periodic testing of contact sentinels or colony animals. Determine a follow-up plan for positives, which will depend on the health status of the room, the types of research, and the tolerance for subclinical infection. Strain and immune status are additional variables to consider when selecting mice for testing. Immune-competent mice should be used for viral serology, as immune-deficient animals will not mount good antibody responses in response to infection. On the other hand, immune-deficient mice are good choices for PCR-based testing, as they tend to be persistently infected with organisms that are quickly eliminated by immune-competent mice. This is particularly important when screening for organisms such as *Pneumocystis murina*, a fungus that is particularly troublesome for severely immune-deficient mice and is typically detected by PCR or histology. For surveillance programs that primarily use colony mice, make sure to test animals from all strains in the room and all investigators. However, if sentinels will be used, some strains are better choices than others. Good choices include BALB/c and DBA, both of which are readily infected with a wide range of pathogens and mount good immune responses once infected. Outbred mice are a reasonable, comparatively low-cost alternative. The final decision is the number of animals to be sampled at each testing interval. The more animals tested, the greater the likelihood of finding a contaminant that has infected only a small percentage of the colony. A common strategy is to test 1–2 dirty bedding sentinels for every 30–50 colony cages, or 5–16 colony animals or contact sentinels per room. The number of animals to test also can be calculated mathematically (Institute for Laboratory Animal Research 1991).

Environmental sampling is increasingly proving to be a viable alternative to sentinel or colony testing for surveillance, although it is most suitable for testing animals housed in individually ventilated cages (IVCs). The theory behind environmental sampling is that genetic material from infectious agents can be found in debris from bedding, which is deposited at various points within the cage and rack exhaust system as air flows through the racks. Benefits of environmental testing include replacement of live animals and increased sensitivity to some infectious agents (Mailhiot et al. 2020; Miller et al. 2016; Zorn et al. 2017) compared with sentinel testing. Environmental sampling uses PCR to identify infectious agents, with several viable options for sampling site, including cage components (Dubelko et al. 2018; Gerwin et al. 2017) and IVC rack exhaust manifolds (typically referred to as exhaust air dust testing, or) (Manuel et al. 2016). Several laboratory animal equipment vendors now offer systems for exhaust air dust testing that integrate with their IVC racks. Cost analysis and parallel testing to compare results with your institution's established methods for infectious agent surveillance are recommended prior to switching to environmental sampling as the sole method of surveillance.

Diagnostic Testing

Health surveillance programs are typically based on batteries or panels of tests. Laboratories that offer comprehensive testing for rodent pathogens often offer defined panels that include both shorter and longer lists of organisms. The specific pathogens considered will depend upon the overall needs of the research facility and the investigator(s). To reduce costs, the shorter panels, which typically include the most commonly seen pathogens, may be selected for most testing periods, with utilization of a longer panel once or twice a year. If using animals, a common diagnostic approach involves a combination of serology for viruses and bacteria, fecal PCR for endoparasites, and cage or fur swab PCR for ectoparasites. Necropsy may also be performed in coordination with the larger panel to include direct microscopic examination for parasites and freezing of tissues such as mesenteric lymph nodes for follow-up testing in the case of an equivocal serologic result. Environmental sampling tends to involve PCR of swabs or cage or rack filter extracts.

disease prevention through sanitation

Practicing proper micro- and macroenvironmental sanitation is an important adjunct to control of infectious disease in animals,

including mice. Cages should be cleaned and disinfected on a regular schedule, although some situations may require more frequent
sanitization and others less frequent. Instruments and equipment
used on more than a single animal should be disinfected between
mice to the greatest practical extent. Optimally, mice infected, or
suspected of being infected, with pathogens should be isolated from
noninfected animals. Further, animals believed to be infected with
pathogens should be handled after other, noninfected animals, and
the traffic pattern in the facility should be designed such that traffic
from areas housing infected animals to those housing noninfected
animals is minimized. If an outbreak is discovered, decontamination
of animal housing systems (cages and racks) and affected vivarium
rooms (walls, floor, ceiling, etc.) should be included as part of the
mitigation plan. Consideration of the environmental hardiness of
the pathogen(s) discovered is important in selecting the appropriate
decontamination method. As agents have been found to be transmitted via feed, movement to irradiated or autoclavable diets is recommended to further eliminate potential sources of contamination
(Adams et al. 2019).

references

Adams SC, Myles MH, Tracey LN, Livingston RS, Schultz CL, Reuter
    JD, Leblanc M. 2019. Effects of pelleting, irradiation, and auto-
    claving of rodent feed on MPV and MNV infectivity. *J Am Assoc
    Lab Anim Sci* **58**:542–550.

Behringer R, et al. 2014. *Manipulating the Mouse Embryo: A Laboratory
    Manual*, 4th ed. New York: Cold Spring Harbor Laboratory Press.

Besselsen DG, Myers EL, Franklin CL, Korte SW, Wagner AM,
    Henderson KS, Weigler BJ. 2008. Transmission probabilities
    of mouse parvovirus 1 to sentinel mice chronically exposed to
    serial dilutions of contaminated bedding. *Comp Med* **58**:140–144.

Compton SR, Homberger FR, Paturzo FX, Clark JM. 2004. Efficacy
    of three microbiological monitoring methods in a ventilated cage
    rack. *Comp Med* **54**:382–392.

Dillehay DL, Lehner ND, Huerkamp MJ. 1990. The effectiveness of
    a microisolator cage system and sentinel mice for controlling
    and detecting MHV and Sendai virus infections. *Lab Anim Sci*
    40:367–370.

Dubelko AR, Zuwannin M, McIntee SC, Livingston RS, Foley PL. 2018. PCR testing of filter material from IVC lids for microbial monitoring of mouse colonies. *J Am Assoc Lab Anim Sci* **57**:477–482.

Gerwin PM, Ricart Arbona RJ, Riedel ER, Henderson KS, Lipman NS. 2017. PCR testing of IVC filter tops as a method for detecting murine pinworms and fur mites. *J Am Assoc Lab Anim Sci* **56**:752–761.

Institute for Laboratory Animal Research. 1991. *Infectious Diseases of Mice and Rats*. Washington, DC: National Academy Press.

Lipman NS, Newcomer CE, Fox JG. 1987. Rederivation of MHV and MEV antibody positive mice by cross-fostering and use of the microisolator caging system. *Lab Anim Sci* **37**:195–199.

Mailhiot D, Ostdiek AM, Luchins KR, Bowers CJ, Theriault BR, Langan GP. 2020. Comparing mouse health monitoring between soiled-bedding sentinel and exhaust air dust surveillance programs. *J Am Assoc Lab Anim Sci* **59**:58–66.

Manuel CA, Hsu CC, Riley LK, Livingston RS. 2008. Soiled-bedding sentinel detection of murine norovirus 4. *J Am Assoc Lab Anim Sci* **47**:31–36.

Manuel CA, Pugazhenthi U, Leszczynski JK. 2016. Surveillance of a ventilated rack system for corynebacterium bovis by sampling exhaust-air manifolds. *J Am Assoc Lab Anim Sci* **55**:58–65.

Miller M, Ritter B, Zorn J, Brielmeier M. 2016. Exhaust air dust monitoring is superior to soiled bedding sentinels for the detection of *Pasteurella pneumotropica* in individually ventilated cage systems. *J Am Assoc Lab Anim Sci* **55**:775–781.

Pilgrim HI, Parks RC. 1968. Foster nursing of germfree mice. *Lab Anim Care* **18**:346–351.

Reetz IC, Wullenweber-Schmidt M, Kraft V, Hedrich HJ. 1988. Rederivation of inbred strains of mice by means of embryo transfer. *Lab Anim Sci* **38**:696–701.

Smith PC, Nucifora M, Reuter JD, Compton SR. 2007. Reliability of soiled bedding transfer for detection of mouse parvovirus and mouse hepatitis virus. *Comp Med* **57**:90–96.

Suzuki H, Yorozu K, Watanabe T, Nakura M, Adachi J. 1996. Rederivation of mice by means of in vitro fertilization and embryo transfer. *Exp Anim* **45**:33–38.

Thigpen JE, Lebetkin EH, Dawes ML, Amyx HL, Caviness GF, Sawyer BA, Blackmore DE. 1989. The use of dirty bedding for detection of murine pathogens in sentinel mice. *Lab Anim Sci* **39**:324–327.

Zorn J, Ritter B, Miller M, Kraus M, Northrup E, Brielmeier M. 2017. Murine norovirus detection in the exhaust air of IVCs is more sensitive than serological analysis of soiled bedding sentinels. *Lab Anim* **51**:301–310.

experimental methodology

restraint

As with any animal, safe handling requires gentleness, firmness, and respect for the animal. People should be thoroughly trained and confident in their ability before they are allowed to handle mice on their own. Fearfulness on the part of the human handler is particularly antithetical to safe handling and must be overcome by practice.

Laboratory mice vary significantly in temperament. Some strains, such as A, are typically placid in response to handling, while others, such as SJL, may be aggressive. Handling wild mice requires focused attention and a gentle, knowledgeable approach, as they are quick, agile, and amazingly adept at escape. Exuberant young mice of 2–4 weeks of age must also be approached with extra care, as they may jump out of the cage as soon as the lid is lifted.

Cage Transfer

To transfer an adult mouse from one cage to another, the mouse may be picked up by the base of the tail (not the tip!) using a gloved hand. Docile mice also can be grasped manually by the scruff of the neck or lifted in a cupped hand. Use of a cup or tunnel to transfer mice is less aversive than tail handling, and mice can be quickly acclimated to these techniques (Gouveia and Hurst 2019; NC3Rs n.d.), although efficiency of cage changing may be sacrificed compared with other methods (Doerning et al. 2019). Blunt or padded-tipped forceps can also be used for cage transfer, although this method has the potential to cause repetitive use injuries (Kerst 2003). In all cases, mice

DOI: 10.1201/9780429353086-6

should never be suspended or dangled and should be placed gently in the new cage, never dropped.

Restraint for Manipulation

If the mouse is to be manipulated in any manner, greater control is needed. To manually restrain a mouse for further manipulation, place it on the cage lid or other rough surface. Pull gently back on the tail, which will encourage the mouse to dig in with all four feet and pull in the opposite direction. Then, quickly and firmly, reach down and grasp the mouse by the scruff near the base of the head (Figure 6.1). With the tail in one hand and the scruff in the other, lift the mouse and tuck the base of the tail between the palm and third or fourth finger of the hand holding the scruff (Figure 6.2). With experience, many handlers learn to perform this entire maneuver in a single step. While effective control necessitates a firm hold on the scruff, make sure that the skin around the neck and chest is not stretched so tightly that it interferes with the animal's ability to breathe. To be able to perform techniques with one's dominant hand, the mouse should be scruffed and restrained using the non-dominant hand.

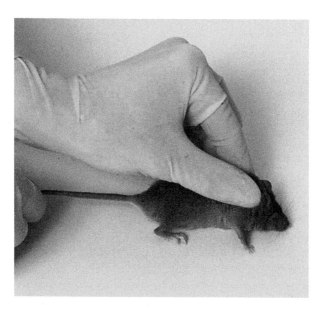

Figure 6.1 Technique for manual restraint of a mouse, Step 1: The mouse is picked up by the scruff of the neck.

Figure 6.2 Technique for manual restraint of a mouse, Step 2: The tail is pinned between the palm and third or fourth finger.

Restraint Devices

Various devices can be obtained commercially or made in the laboratory to restrain mice for further manipulations. For quick procedures, these devices seldom offer any advantage over manual restraint, but they can be quite useful for procedures that require more prolonged restraint or repeated manipulations at short intervals. Prolonged restraint should be scientifically justified by the investigator and approved by the Institutional Animal Care and Use Committee (IACUC). Refer to your institution's IACUC policies and guidelines for details on what constitutes prolonged restraint.

sampling methods

Blood

The blood volume of the average adult mouse is only 2–2.75 ml, and the peripheral veins are small. This means that blood samples must be comparatively small and may be challenging to obtain without harming the mouse. The following are some guidelines regarding sample size.

1. Approximately 1% of the body weight can be removed safely at one time without fluid replacement (0.25 ml from a 25-g mouse; 0.35 ml from a 35-g mouse).

2. If the total volume of a single draw is over 1% (not to exceed 2%) of body weight, fluid volume must be replaced through administration of an isotonic sterile fluid solution such as saline. Ideally, fluids should be given prior to blood collection to prevent physiologic alterations associated with acute blood loss (Marx et al. 2015) (0.35–0.5 ml from a 25-g mouse; 0.5–0.7 ml from a 35-g mouse).

3. For repeated sampling, if less than 1% of the body weight is taken in total over a 14-day period, no additional action is necessary. If over 1% of body weight in total is taken over a 14-day period, fluid replacement is necessary. The hematocrit content, hemoglobin content, or both should be monitored in all animals that will have large volumes (i.e., approaching the recommended maximum) drawn repeatedly (more than three times).

The following are some of the more common non-terminal (1-6) and terminal procedures (7-8) used to obtain blood samples from mice.

1. **Facial vein (aka "cheek bleed")**: The mouse is restrained, typically on its side on a flat surface. Using a lancet or 22–25-gauge needle held perpendicularly, with the tip facing slightly toward the nose, firm pressure is applied to pierce the skin at a point caudodorsal to the sebaceous gland, a small, hairless "dimple" found caudal to the corner of the mouth slightly below the jaw line (Figure 6.3). The placement of the lancet or needle perpendicularly is key to avoiding puncture of the ear canal. Once the appropriate volume of blood has been collected, typically in a collection tube held just below the lateral aspect of the jaw (Figure 6.4), pressure is applied with sterile gauze to the site to stop the bleeding. Serial blood collection from this site has been shown to cause serious adverse clinical events, including mortality, convulsions, head tilt, and hemorrhage from the ear canal and nares (Frohlich et al. 2018). Those using this technique must be well trained and proficient to avoid complications. Expected collection volume using this technique is 200–500 µl.

2. **Lateral tail vein**: The mouse has two lateral tail veins and a ventral tail artery. The lateral tail vein can be punctured to allow collection of up to a few drops of blood. The mouse is restrained on a flat surface or placed in a restraint device

Figure 6.3 Technique for blood collection from the facial vein, Step 1: Using a specially designed lancet or 25-gauge needle, a stab incision is made into the cheek caudodorsal to the cheek skin gland.

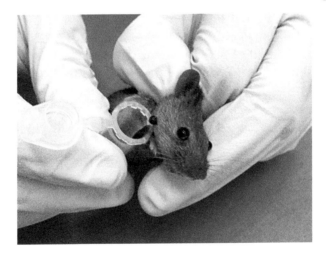

Figure 6.4 Technique for blood collection from the facial vein, Step 2: Drops of blood are caught in a collection tube held just below the lateral aspect of the jaw.

with the tail protruding. The tail is disinfected with 70% alcohol and allowed to dry. The veins at the base of the tail can be visualized and punctured approximately 2–3 cm from the tail tip with a 22–25-gauge needle. Blood can be drawn into a microhematocrit tube or slowly aspirated using a 0.5–1.0-ml syringe. Warming the animal or its tail beforehand will dilate the vessels, improving puncture accuracy and possibly increasing yield. Consideration should be given to the use of tail vein laceration versus puncture. There is the potential that the tail artery will be accidentally lacerated, resulting in severe hemorrhage. In addition, laceration causes additional tissue damage compared with puncture and can therefore be more painful and take longer to heal. Expected collection volume using this technique is 50–100 μl.

3. **Distal tail transection (1–3 mm)**: After disinfection of the tail, a single perpendicular complete transection is made at the tip of the tail with a scalpel. Special care must be taken so that the transection does not include tail vertebrae, *only* soft tissue. Blood is collected from the tail tip. For serial collection over a short period of time, the scab is removed for each additional sample. This technique is useful for repeated blood collections over the course of several hours, such as for a glucose or insulin tolerance test. At least one study suggests this method has less impact on animal welfare than the facial vein technique (Moore et al. 2017). Consult with your institution's IACUC for requirements for maximum transection length and use of analgesics or anesthetics with this technique. Expected collection volume using this technique is <100 μl.

4. **Saphenous vein**: The lateral saphenous vein runs dorsally and then laterally over the ankle of the hind leg. To access it, the mouse is placed headfirst in a plastic tube. With one hand maintaining a firm hold on the tail, the other hand is used to extend the hind leg and immobilize it in an extended position by applying gentle downward pressure immediately above the knee joint (Figure 6.5). To access the medial saphenous vein, which runs along the medial side of the leg, restrain the animal with the non-dominant hand. Use the dominant hand to extend the hind leg and immobilize it against the palm of the restraining hand by holding the foot down with the pinkie finger. Once the animal is properly restrained, the lateral or medial saphenous vein is easily seen after fur is clipped from the site or the fur is wetted with alcohol (Figure 6.6). When the vein is punctured with a 25-gauge needle, a drop of blood

Figure 6.5 Technique for blood collection from the lateral saphenous vein. The mouse is restrained in a plastic tube, the leg is shaved over the knee and wiped with 70% ethanol, and the vein is punctured with a 25-gauge needle.

Figure 6.6 The medial saphenous vein, visible after the inner side of the leg is shaved and wiped with 70% ethanol. For blood collection, the vein is punctured with a 25-gauge needle.

should appear immediately at the puncture site. After collection is completed, stop the bleeding by releasing the leg and, if necessary, applying gentle pressure to the puncture site. Expected collection volume using this technique is 100–200 μl.

5. **Retro-orbital sinus puncture**: The mouse should be anesthetized for this procedure unless withholding of anesthesia is scientifically justified. The mouse is manually restrained, and a microhematocrit tube or small-bore pipette is placed at the medial or lateral canthus of the eye. The tube is rotated and directed caudally at a 30° angle. As the sinus is ruptured, blood will flow back into and through the tube (Figure 6.7). After the tube is withdrawn, gentle pressure should be applied with cotton or a gauze sponge to stop the bleeding. If personnel are properly trained and skilled in this technique, it can be accomplished with little trauma and will be followed by rapid healing. The risk of permanent damage increases each time the procedure is performed, and so alternate eyes should

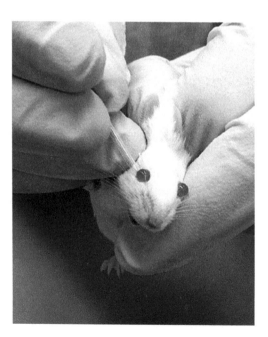

Figure 6.7 Blood collection from the retro-orbital sinus. A glass capillary tube is inserted into the lateral (shown) or medial canthus and gently rotated until the sinus is penetrated and blood flows into the tube.

be used for serial sampling, and at least 2 weeks should be allowed between sampling from the same site. Expected collection volume using this technique is 200 μl.

6. **Other non-terminal blood collection techniques**: Several studies have been published describing alternative sites for blood collection in the mouse, including the sublingual vein (Heimann et al. 2009, 2010), dorsal pedal vein (Parasuraman et al. 2010), and chin bleed (submental vein or inferior labial vein) (Constantinescu and Duffee 2017; Regan et al. 2016). As of publication of this edition, these methods have not been widely adopted, and their use should be justified and approved by your institution's IACUC.

7. **Cardiac puncture**: This should be done only as a terminal procedure in an anesthetized mouse. The animal is placed on its back on a flat surface. A 25–22-gauge needle is either: (1) inserted through the diaphragm lateral to the xiphoid cartilage and directed forward and medially toward the heart (Figure 6.8); or (2) inserted between the fifth and sixth ribs on the left side and directed forward toward the heart. This method requires training and skill, but can be used to obtain comparatively large volumes of blood. Expected collection volume using this technique is 1–2 ml of blood.

8. **Decapitation**: This is another technique that can be used to obtain comparatively large volumes of blood, provided that contamination of the blood with hair and body fluids is not contraindicated. Personnel must be well trained to perform this method safely, and it should be used only when scientifically justified by the investigator. Decapitation may cause significant pain for the mouse if performed incorrectly. The potential for pain is reduced or eliminated if the mouse is anesthetized. Expected collection volume using this technique is 1–2 ml.

For serial sampling from sites with alternate sides (left and right), collection sites should be alternated each time to avoid excessive tissue disruption.

Urine

Mice will often urinate upon handling, and, if a test tube is readily at hand, a drop or two of urine can be caught from the genital

Figure 6.8 Technique for blood collection by cardiac puncture. A 24-gauge needle is inserted through the diaphragm lateral to the xiphoid cartilage and directed forward and medially into the cardiac ventricle. Diaphragm, lungs, and heart are illustrated using dotted lines.

papilla immediately after the mouse has been lifted from the cage. If necessary, the mouse can be restrained manually, and the caudal abdomen (over the bladder) gently massaged to stimulate urination. The catch tube should be prepositioned beneath the genital papilla to catch the urine.

If more than a drop or two of urine is needed, it may be necessary to make multiple collections from the mouse. Alternatively, metabolic cages or hydrophobic sand can be used to collect larger volumes of urine. Metabolic cages typically consist of a wire-bottom floor on top of a funnel-shaped base. Urine (and feces) drop through the wire floor and are funneled into a collection vessel placed underneath the cage. Samples obtained in this manner are not useful for microbiological analysis because of mixing of sample types. Hydrophobic sand, originally developed for urine collection in cats, has also proven successful for urine collection in rodents and could be less stressful to mice than placement in a metabolic cage (Hoffman et al. 2017, 2018, 2019). Currently marketed as "LabSand" to the scientific community, hydrophobic sand is biodegradable and non-toxic. Its hydrophobic properties cause urine to pool on the surface, allowing for easy collection.

Finally, several publications describe techniques for bladder catheterization in female (Reis et al. 2011; St Clair et al. 1999) and male (Lamanna et al. 2020) mice, although anesthesia is required.

Feces

Dry, and sometimes fresh, feces can be obtained from the cage bottom. To obtain a fresh sample from the mouse, remove the animal from its cage and set it on the cage top or other surface. Place a test tube beneath the anus and gently press it up and back to "milk" a fecal pellet out of the rectum and into the tube. With some mice, it may be possible to obtain two or three pellets within a short period of time by placing the mouse back in the cage after collecting a pellet, then coming back and repeating the process a few minutes later. Alternatively, mice can be placed in an empty cage, where they will inevitably defecate, allowing fresh fecal samples to be collected from the cage floor. Feces also can be collected using a metabolic cage.

Samples for DNA Analysis

Genetic testing is commonly performed in modern mouse colonies to determine which mice carry a desired genetic modification. Genotyping is done by analyzing a tissue sample from the mouse by Southern blotting or PCR (Chatelain et al. 1995; Sato et al. 1995). A variety of tissues have been used for this purpose, including blood (Winberg 1991), saliva (Irwin et al. 1996), ear punch biopsies (Garzel et al. 2010; Ren et al. 2001), buccal swabs (Meldgaard et al. 2004), feces (Broome et al. 1999; Kalippke et al. 2009), hair (Schmitteckert et al. 1999), and rectal epithelial cells (Lahm et al. 1998). The most commonly used tissue, however, is the tip of the tail. When performed on young mice before weaning, excision of tissue from the tail tip is typically performed without anesthesia. The mouse is restrained, and a very small (about 2-mm) section is removed from the end of the tail using a single stroke with a scalpel blade. Pressure over the wound may be needed to stop bleeding. With older mice, or if more tissue will be removed, local or general anesthesia should be used. Immersion of the tail in ice-cold 70% ethanol for 10 seconds is a simple but effective method of topical anesthesia for tail biopsies (Dudley et al. 2016). Because of potential prolonged inflammatory changes associated with topical vapocoolant anesthesia, this method should be avoided (Braden et al. 2015). Bleeding may also be more of a problem with older mice, and the wound will generally take longer to heal. Excision size and age limits are based largely on the fact that, once ossification of tail vertebrae is complete, there is additional pain associated with tail biopsy because of the incorporation of bone into the excision site. In multiple strains, ossification is complete by 17 days of age, and vertebrae extend up to 2 mm from the distal tail tip (Hankenson et al. 2008). Therefore, to the extent possible, tail excision for genotyping should be performed prior to 17 days of age, with excision of the minimum length of tail necessary.

Vaginal Swabs

Vaginal swabs can be used to identify the stage of the estrous cycle in a female mouse. Although of limited practical utility for breeding of mice, this technique is sometimes employed for experimental purposes. To obtain a swab, the mouse is restrained manually, and a moistened cotton-tipped swab is inserted into the vagina (be aware of the size of the swab to ensure that the collection technique does not cause tissue trauma—an appropriately sized swab can be made using a bit of sterile cotton wool wrapped around a blunt toothpick). The swab is gently but firmly rotated, then removed and wiped onto a clean glass microscope slide. Once the slide has air dried, it is stained with a 0.1% aqueous solution of methylene blue. After the stain has dried, the slide is examined under a microscope and interpreted as follows:

- **Diestrus**: Cells consist primarily of polymorphonuclear lymphocytes (PMNs), with some epithelial cells.
- **Proestrus**: Nucleated and cornified epithelial cells, with some PMNs in the early stage.
- **Estrus**: Cornified epithelial cells predominate, with a few nucleated cells seen in the early stage.
- **Metestrus**: Cornified epithelial cells and PMNs predominate, with some nucleated epithelial cells.

Images of vaginal cytology and additional information about the mouse estrous cycle are available in Byers et al. (2012).

compound administration

Drugs and other substances can be administered to laboratory mice by a variety of routes. People using these techniques should be thoroughly trained and skilled before working with live, unanesthetized animals. Common routes of compound administration include the following:

Oral (PO)

Some compounds can be administered in the food or water; however, it is difficult to give a precise amount or dose in this manner, and mice may be reluctant to consume the treated food or water at all. For this reason, direct administration by oral gavage is often the preferred approach. For this procedure, the distance from the mouse's nose to the end of the ribcage is measured and marked on the gavage needle to ensure that the needle is not passed through the stomach. The mouse is manually restrained with neck extended—firm restraint with complete immobilization of the mouse's head is essential for this procedure—and a gavage needle is passed through the animal's mouth and into the esophagus or stomach, being certain not to go past the marking on the needle (Figure 6.9). The compound is administered slowly using a syringe attached to the gavage needle. Successful gavage should not require significant amounts of force to progress the gavage needle—consistent gentle pressure should be sufficient. The primary complication that may occur using this technique is passage of the needle into the trachea or lungs rather than the stomach. Signs that this may have occurred include choking, coughing, vigorous struggling, fluid emanating from the nose or mouth, or sudden death. Any of these signs should prompt immediate withdrawal of the needle. In addition, traumatic rupture of the pharynx or esophagus may occur if excessive force is used in passing the gavage needle. Upon injection of the sample, hydrothorax will result if these tissues have been ruptured. This can also manifest as subcutaneous emphysema or swelling in the neck/shoulder area as food material accumulates in this space. If it appears that fluid has gotten into the lungs, or the pharynx or esophagus has been

Figure 6.9 Orogastric gavage. The tip of the feeding needle is directed to the back of the pharynx, through the esophagus, and into the stomach. Care must be taken not to enter the trachea.

ruptured, the mouse should be euthanized immediately. Sedation or anesthesia prior to gavage is not recommended as this removes the animal's ability to protect its airway and increases chances for intratracheal administration. Volume administered via oral gavage should be limited to 5–10 ml/kg.

Intramuscular (IM)

The intramuscular route is not used as often in mice as in larger species because of the small muscle mass of the mouse. In mice, intramuscular injections are generally made into the muscles at the back or top of the thigh. The mouse is restrained manually, and the tip of a 25-gauge needle (attached to a syringe) is inserted firmly through the skin and into the muscle. Prior to administration of the compound, it is advisable to pull back lightly on the plunger of the syringe. If blood is pulled back into the syringe, the needle is probably in a blood vessel and should be discarded. Volume administered via IM route in mice should be limited to 50 μl or less to prevent distribution into extramuscular tissues (Gehling et al. 2018).

Intraperitoneal (IP)

The mouse is restrained manually and held with the head and body tilted slightly downward. With a firm motion at a 30–45° angle from the mouse's body, the tip of the needle is inserted into the caudal abdomen through the skin and just through the abdominal wall (Figure 6.10). The insertion is made to either the right or left of the midline (to avoid the cecum, which typically lies at, or slightly to the side of, the midline in mice), in the lower quadrant of the abdomen. Prior to administration of the compound, the plunger of the syringe should be pulled back slightly. If any fluid (urine, blood, intestinal contents) is seen in the syringe or needle hub, the needle is probably in one of the abdominal organs and should be discarded. With training and practice, it is possible to achieve a high rate of success with IP injections. However, there is always a risk of injection into an abdominal organ. This may have no serious consequences for the mouse, but may be associated with slow and erratic absorption of the

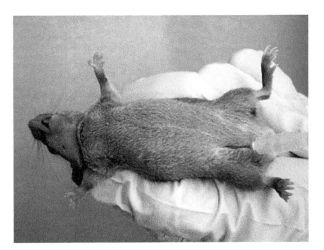

Figure 6.10 Intraperitoneal injection. With a firm motion, the tip of the needle is inserted into the caudal abdomen, to the right or left of the midline, through the skin and just through the abdominal wall. Before injection, the plunger of the syringe is pulled back to assure that neither blood vessels nor internal organs have been penetrated. Volume administered via IP route in mice should be limited to 10 ml/kg.

compound. This is a particular problem in the case of anesthetics, as the mouse will not respond as expected to the drug. Re-dosing under these circumstances is tricky, as the initial dose will eventually be absorbed and, combined with the second dose, may prove fatal.

Subcutaneous (SC)

Depending on how the mouse is restrained, SC injections may be made through the loose skin over the upper back and neck or through the considerably tighter skin over the ventral abdomen. The tip of the needle is inserted firmly through the skin, and the needle is advanced several millimeters farther. No resistance should be encountered as the plunger of the syringe is depressed. Resistance suggests that the needle has not completely penetrated the skin. Volume administered via SC route in mice should be limited to 40 ml/kg.

Intradermal (ID)

Intradermal injections are made into the skin over the back/neck, ventral abdomen, or hind footpad. The technique is much the same as for SC injections except that the tip of the needle is placed between the layers of skin rather than through the skin. In contrast to SC injections, resistance should be felt both as the needle is advanced and as the compound is injected. A hard bleb will be seen upon successful ID injection of even a small quantity of fluid. This is a tricky procedure in any species, but is particularly difficult in the mouse because the skin is so thin. Removal of the hair overlying the injection site will facilitate visualization and improve the likelihood of success. Volume administered via ID route in mice should be limited to 0.05–0.1 ml per injection.

Intravenous (IV)

The most common sites for IV injections in the mouse are the lateral tail veins. These vessels are readily visualized but are quite small in diameter, and injection into them requires considerable practice and skill. The mouse is typically placed in a restraint device for this procedure. Warming the mouse, or just its tail, will dilate the veins and make the procedure easier. Excellent lighting will also prove helpful. A small-gauge (27-gauge or smaller) needle is inserted just through the skin and into the vein; the tip is then advanced a couple of millimeters (Figures 6.11 and 6.12). If placement has been successful, there will be no resistance felt as the compound is administered, and the vein will appear to "bleach out" as the administered fluid temporarily replaces the blood within it. Volume administered via IV route in mice should be limited to 5 ml/kg as a bolus, or 4 ml/kg/h as a constant infusion.

Retro-orbital (RO)

The mouse should be anesthetized for this procedure, which is suitable for administration of very small quantities of fluid (Yardeni et al. 2011). It may also be used to inject cells into the bloodstream, but this is not advised since cells may engraft behind the eye, causing welfare concerns. With the mouse held firmly in one hand, skin at the side of the animal's face is stretched back to expose the membrane of the retro-orbital sinus. A small (26-gauge or smaller) needle is inserted through the conjunctiva of the eye at the medial or lateral canthus until the bevel is fully in the sinus, after which the plunger of the syringe is depressed. Volume administered via RO route in mice should be limited to 0.05 ml total.

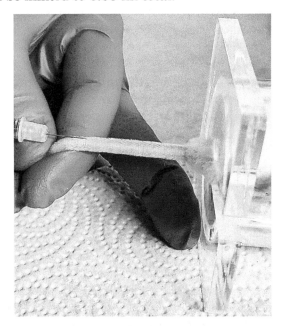

Figure 6.11 Intravenous injection. The mouse is placed in a restraint device to facilitate injection into one of the lateral tail veins.

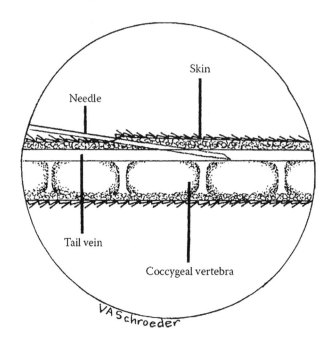

Figure 6.12 Illustration of the internal structures of the tail during tail vein injection. Note the penetration of the needle through the skin and into the vein.

Implantable Cannulas and Pumps

Osmotic pumps can be placed subcutaneously or into the abdominal cavity to deliver compounds at a slow, steady rate over a period of days or weeks. Implantation is a surgical procedure and should be performed using aseptic technique. Implantable cannulas permit continuous access to the venous or arterial system for either IV compound administration or blood withdrawal. This technology is used more often in larger animals, but miniaturized devices for mice are commercially available. Using strict aseptic technique, the cannula is inserted into a vein or artery (the femoral vessels, jugular vein, and carotid artery are common sites) and secured in place. The other end of the cannula is tunneled under the skin to a location between the shoulders, where it may be: (1) exteriorized through the skin, anchored to the muscle, and capped with a heparinized lock and stainless steel pin (an externalized cannula is accessed for injections or blood withdrawal by removing the pin); or (2) attached to a small port that is secured to underlying tissues beneath the skin. (To access a subcutaneous port, the skin over it must be shaved and disinfected. Special Huber point needles may be used to prolong the life of the port.) More information on implantable cannulas can be found elsewhere (Desjardins 1986; Hayward et al. 2007).

anesthesia and analgesia

Anesthesia

Anesthesia can refer to general anesthesia, in which conscious awareness is lost, or local anesthesia, in which sensation is lost in a part or parts of the body, but consciousness is maintained. The goal of anesthesia is to eliminate the perception of pain, to immobilize the animal to permit some manipulation, or both. These goals must be accomplished without posing a serious threat to the life of the animal. In addition, to be useful in a research setting, the anesthetic must have as little effect as possible on the parameter(s) being studied. No single anesthetic or combination of drugs or techniques is the best choice for all situations. The key to choosing a method of anesthesia is to give careful thought beforehand both to what you want the anesthetic to do and what you do not want it to do. Consultation with a veterinarian can then help you decide what method is likely to work the best. It is important to remember that no anesthetic is both 100% effective and absolutely safe.

The following variables should be considered when selecting a method of anesthesia:

1. **The animal**: Mice vary significantly in their sensitivity to both the desired and undesired effects of anesthetics. Age, body composition, health status, and sex may all affect the response to an anesthetic, and remarkable variability may be seen in the responses of mice of different stocks/strains.

2. **The procedure**: Anesthetics vary in their duration of effect and degree of analgesia provided. Thus, consideration should be given to the length of the procedure for which the animal will be anesthetized and the amount of pain that might result.

3. **The laboratory**: In this sense, the laboratory is the environment in which the animal will be anesthetized. What kind of equipment is available for anesthetic administration (e.g., precision anesthetic vaporizer)? Will someone be able to monitor the condition of the anesthetized animal? Is there

any equipment available for monitoring the condition of the animal? Will someone be available who knows what to do if the animal fails to respond satisfactorily to the anesthetic or experiences a severe adverse reaction? If the answer to any of these questions is "no," then it might be preferable to choose a drug with a wide margin of safety.

4. **Post-procedural use of the animal**: Most drugs will have some kind of lingering effect on the animal that may or may not be consistent with post-procedural care capabilities or the animal's intended use in research. For example, some drugs may have a prolonged depressant effect on respiration, temperature regulation, or both. If the mouse cannot be monitored and kept warm after the procedure, the post-procedural mortality rate might be high with these drugs. If it will be difficult to monitor the animal closely after the procedure, it would be desirable to choose a drug that has a low potential for a long or difficult recovery from anesthesia. Alternatively, a drug could be chosen that can be reversed by administration of a specific antagonist.

Methods and drugs

There are many established protocols for mouse anesthesia. The reader is referred to an excellent review of this topic by Gaertner et al. (Gaertner et al. 2008). Some of the more commonly used compounds for anesthesia of mice are described and summarized, with doses, in Table 6.1. (*Note*: The doses in Table 6.1 are starting doses,

TABLE 6.1: COMMONLY USED COMPOUNDS FOR GENERAL ANESTHESIA OF MICE[a]

Anesthetic	Dosage	Route of Administration
Isoflurane	1–5% (to effect)	Inhalation
Pentobarbital	50–90 mg/kg (diluted 1:9 in sterile saline)	IP
Ketamine + xylazine	80–100 mg/kg ketamine + 7.5–16 mg/kg xylazine	IP, IM, SC (Levin-Arama et al. 2016)
Ketamine + xylazine + acepromazine	80 mg/kg ketamine + 8 mg/kg xylazine + 1 mg/kg acepromazine	IP or IM
Ketamine + medetomidine	75 mg/kg ketamine + 1 mg/kg medetomidine	IP

[a] The doses given are approximate starting doses; re-dosing may be required. Intraperitoneal continuous rate infusion (CRI) has been used successfully with some drug combinations (Erickson et al. 2016).

and additional drug may be needed to achieve the desired plane of anesthesia.)

Inhalant anesthetics

Inhalant anesthetics are delivered to the patient as gases or vapors. In general, these agents act by depressing transmission of nerve impulses in the brain, which results in a blocking of the transmission of messages from the brain to the rest of the body. This results in a loss of motor control, depression of the cerebral cortex, and loss of consciousness.

The most commonly used inhalant anesthetic for mice is isoflurane. Advantages of this drug include the following:

- Rapid induction and recovery;
- Comparatively little effect on cerebral blood flow and cerebrospinal fluid pressure; pronounced neuroprotective effect;
- Minimal effects on coronary blood flow and cardiac output; protective effect in models of ischemic cardiac disease;
- Minimally metabolized by the liver, with little effect on hepatic microsomal enzymes;
- Good muscle relaxation.

Disadvantages of isoflurane include the following:

- Expensive;
- Pungent odor (can cause animals to hold breath during induction);
- Increased airway secretions and reflexes;
- Requires vigilant monitoring, since the depth of anesthesia may change rapidly.

With the proper vaporizer and scavenging equipment, isoflurane is probably the most desirable anesthetic for mice, especially for long procedures. It is important that the mouse be monitored carefully to avoid overdose, however.

Initial anesthesia of mice with gaseous anesthetics is generally accomplished by placing the animal in an induction chamber. This is a small chamber with a tight-fitting lid. Anesthetic in the appropriate concentration is piped into the chamber from the anesthetic vaporizer (Figure 6.13). Once the mouse is anesthetized, it is removed from

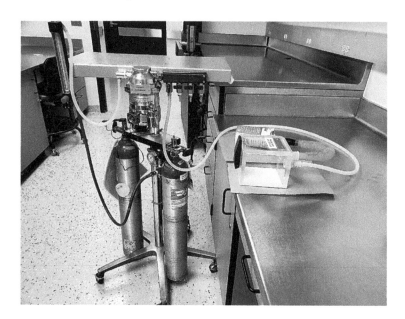

Figure 6.13 A typical setup for inhalation anesthesia of mice with a commercially available precision vaporizer and induction chamber.

the chamber, and anesthesia is maintained by delivery of the gas through a face mask, which should fit as snugly as possible around the nose of the mouse (Figure 6.14). It is also possible to deliver anesthetic to a mouse through a tube inserted into the trachea (Hamacher et al. 2008; Spoelstra et al. 2007; Watanabe et al. 2009). Whenever inhalant anesthetics are used, it is essential that appropriate measures be taken to prevent exposure of human personnel to anesthetic vapors. This may be accomplished by performing the procedure in a properly ventilated fume hood or by using any of a variety of scavenging systems (Vogler 2008).

Injectable anesthetics

Barbiturates: The barbiturate anesthetics are injectable drugs that appear to interact with specific receptors in the central nervous system to inhibit the transmission of nerve impulses. They can be used to produce all levels of clinical depression, from hypnosis to coma. The following are general characteristics of barbiturates:

- They have a relatively narrow margin of safety, with wide variability in the effective and lethal doses. Mortality with light anesthetic doses is possible.

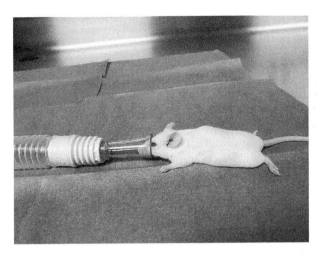

Figure 6.14 Use of a face mask for maintaining anesthesia following induction.

- All cause respiratory depression, which increases with the dose of the drug and may be severe in unborn pups (if the drug is administered to the pregnant dam). They also cause disturbances in respiratory rhythm. Complete cessation of breathing (apnea) occurs commonly, especially after intravenous administration.
- All cause cardiac depression, increases in cerebral blood flow and intracranial pressure, and possible cardiac arrhythmias.
- They may cause excitement during induction, especially if the drug is injected too slowly.
- They have no analgesic effect and may actually increase the perception of pain when used at subanesthetic doses (Marshall and Longnecker 1996).
- Muscle relaxation is moderate.
- They interfere with the regulation of body temperature, which may lead to significant hypothermia.
- They must be administered IV or IP. Administration by the SC or IM routes will result in significant irritation or even necrosis at the site of injection.

Pentobarbital is the most common barbiturate used in mouse anesthesia. It is a short-acting barbiturate, with a dose-dependent duration of effect. High doses provide 20–45 min of light surgical anesthesia. There is considerable variation in response to this drug related to mouse strain, age, sex, bedding, environmental temperature, and nutritional status.

Barbiturates have many advantages that made them popular for mouse anesthesia in the past. They are readily available, comparatively inexpensive, and generally suitable for administration by the intraperitoneal route. There is also a large body of data related to their effects on various experimental parameters. However, they also have many disadvantages, and for most situations today, there are better choices.

Ketamine hydrochloride and its combinations: Ketamine induces a state of dissociative anesthesia, in which the subject experiences an altered state of consciousness without complete loss of consciousness. In mice, this state is characterized by the following:

- Immobility associated with increased muscle tone and sudden, jerky movements;
- Variable analgesia, which is generally inadequate for surgery in mice;
- General stability of respiratory function;
- Stimulation of most cardiovascular parameters with increased cerebral blood flow, intracranial pressure, and intraocular pressure.

Ketamine alone is not useful for anesthesia of mice.

Muscle relaxation can be greatly improved and the duration of anesthesia prolonged by combining ketamine with sedative drugs such as diazepam or acepromazine. The degree of analgesia with these combinations is inadequate for surgery in mice, but they provide good restraint for mildly painful, nonsurgical manipulations.

For moderate surgical anesthesia, ketamine can be combined with an alpha-2 agonist such as xylazine or medetomidine. In addition to better muscle relaxation and analgesia, these combinations cause moderate to severe hypothermia and respiratory and cardiovascular depression. These effects, along with the analgesia and muscle relaxation, can be reversed with atipamezole (Janssen et al. 2017). Note that, when ketamine is used in combination with alpha-2 agonists, the alpha-2 agonists should not be re-administered if the animal fails to reach an appropriate plane of anesthesia. Because of the hemodynamic effects of alpha-2 agonists, re-dosing can cause significant physiological alterations that can lead to anesthetic complications and death. Rather, ketamine alone should be re-administered at ¼–⅓ the original dose. If ketamine and xylazine are used, and there is consistent difficulty reaching or maintaining a surgical plane of anesthesia, the sedative acepromazine can be added to enhance anesthesia and provide a more reliable surgical anesthetic plane. If issues with anesthetic depth persist, contact your institution's veterinarian for assistance.

Tribromoethanol: Because it is not available as a pharmaceutical-grade compound, and because safer alternatives are available, use of tribromoethanol for anesthesia is no longer recommended and requires approval from an institution's oversight body (i.e., IACUC).

In the past, tribromoethanol was sold commercially in the United States under the brand name Avertin®. Today, it is available as a nonpharmaceutical-grade white, crystalline powder that is typically diluted in vehicle (butanol, amylene hydrate, or amyl alcohol) to form a stock solution. The stock solution is diluted in distilled water or saline at 40–50°C to form a working solution. It should be filtered prior to use. Characteristics of tribromoethanol include the following:

- Rapid induction with 15–20 min of moderate surgical anesthesia;
- Good muscle relaxation, fair analgesia;
- Moderate cardiovascular and respiratory depression at moderate anesthetic doses (severe at high doses).

The breakdown products of tribromoethanol are highly irritating to tissues. Following IP administration of the drug, these breakdown products can cause adhesions, peritonitis, intestinal disorders (including ileus), and death (Gaertner et al. 2008; Meyer and Fish 2008; Papaioannou and Fox 1993). These effects are most likely to occur in these cases:

- If the solution is old or was prepared or stored improperly. It should be stored in a cool, dark place and used within 2–3 weeks of preparation.
- Following the second exposure to the drug, regardless of the dosing interval.

If properly prepared and stored, tribromoethanol is a reliable anesthetic for mildly to moderately painful procedures of short duration. Supplemental administration of the drug to prolong anesthesia or repeated use of the drug in individual animals for recovery procedures can increase the potential for adverse effects.

Special techniques

Neonates: Neonatal mice are difficult to anesthetize safely. The margin of safety with most anesthetics is much narrower for neonates than adults, and the mortality rate associated with dosages adequate to produce surgical anesthesia can be quite high. Nonetheless, neonatal mice appear to be at least as capable as adults of perceiving pain and must be anesthetized for painful procedures. The following techniques can be used to anesthetize young mouse pups.

1. **Hypothermia**: Anesthesia and analgesia result from cold-induced depression of neural conduction and synaptic transmission, which completely cease at 9°C. Neonatal mice cool rapidly and tolerate cooling down to 1°C. Profound hypothermia can be safely maintained for up to 30 min. Hypothermia is sufficient for surgery in mouse pups up to 6–7 days of age. To prevent pain during cooling, pups should not be placed in direct contact with a freezing or frozen surface.

2. **Isoflurane**: Isoflurane is safe and effective for anesthesia of neonatal mice. As with anesthesia of adults, it should be used with proper delivery and scavenging equipment.

Several other agents, including ketamine and pentobarbital, are associated with high mortality rates in neonatal altricial rodents (Danneman and Mandrell 1997) and are not recommended for use with young mouse pups.

Periprocedural Care

If a mouse is to be anesthetized for any reason, it is important that the animal be given special attention before, during, and after the procedure. All anesthetics have the potential to significantly alter the animal's physiologic status, and, without proper care, the outcome could be fatal. The type and severity of physiological alterations will vary with the anesthetic chosen, but most anesthetics will cause at least some degree of respiratory and cardiovascular depression, and virtually all will cause some degree of hypothermia. These changes will persist and may even worsen until the animal has recovered fully from anesthesia, and so supportive care and monitoring may be necessary for many hours after the procedure has ended. As with adverse physiological effects, the time to complete recovery will vary with the anesthetic.

The ideal candidate for anesthesia is a normal, healthy mouse. The type of preoperative evaluation and blood work that is typically performed on larger animals is seldom done with mice. However, the mouse should look active and alert, with bright eyes and a well-groomed appearance. Realistically, this standard cannot always be met. Many mutant and genetically engineered mice will not appear normal, and many of these mice are less-than-ideal candidates for anesthesia. Successful anesthesia of such mice can be tricky, and it is particularly important that they be given careful attention during and after the procedure. The anesthetic chosen for these animals should be one that offers a wide margin of safety, one that can be terminated or reversed if the mouse experiences severe adverse effects, or both.

Intraprocedural monitoring involves two aspects: (1) assuring adequate depth of anesthesia for the type of procedure to be performed and (2) evaluation of the physiological status of the anesthetized animal. For anesthetic depth monitoring, the most reliable indicator is the response to toe pinch; the response to tail pinch is often used but is a less reliable indicator. If the mouse moves or increases its rate of breathing in response to the pinch, it is inadequately anesthetized for a painful procedure. Note that the corneal and palpebral reflexes, which are commonly used to assess the depth of anesthesia in larger animals, are unreliable in mice.

The requirement for physiological monitoring and support will depend on the procedure itself, the anesthetic used, and the requirements set forth by each institution's IACUC. As a general rule, physiologic monitoring and support should be provided for any procedure lasting 30 min or more. In terms of the type of monitoring performed, at the very least, the mouse should be checked frequently to ensure that it is still breathing regularly. More sophisticated monitoring techniques are also available for use in mice, including techniques for monitoring blood oxygen saturation, blood pressure, and electrical activity of the heart.

Basic physiologic support involves temperature and blood pressure maintenance. Temperature maintenance can be done with a homeothermic heating pad, which incorporates a feedback mechanism to adjust heating pad settings based on the temperature of the mouse, a standard heating pad, or a recirculating warm water blanket. Homeothermic heating pads and recirculating warm water blankets are the most likely to provide adequate thermal support with minimal risk of thermal burns. Care must be taken with other types of heating pads to guard against thermal injury to the mouse. If electric heating pads are used, they should be checked for "hot spots" prior to placing the animal on them. Heat lamps are generally discouraged since there is an increased risk of thermal injury. It is easier to maintain body temperature than to rewarm a cold animal, and so heat support should be provided as soon as anesthesia is induced. Blood pressure support can easily be accomplished in uncomplicated procedures through administration of subcutaneous fluids after anesthetic induction. As most anesthetic agents decrease blood pressure, providing additional blood volume through fluid administration can prevent complications associated with hypotension. In general, 0.5–1 ml of isotonic sterile fluids is appropriate for procedures on adult mice lasting up to 2–3 h, with repeat administration for prolonged procedures. If blood loss occurs during the procedure, additional fluids should be given equal to the estimated volume of blood loss. To prevent hypothermia associated with fluid administration, fluids should be warmed prior to administration. Finally, for all but very brief periods of anesthesia, use of a lubricant for the eyes is recommended to prevent corneal injury associated with loss of blink reflex under anesthesia. Ophthalmic ointments for this purpose are commercially available.

Monitoring and physiologic support should continue after the procedure has ended. Until the mouse is up and moving about, it should be checked frequently to ensure that it is breathing normally. It should also be kept warm, but care must be taken to avoid thermal injury or overheating. The best option is to place the animal's cage on a heating pad in such a manner that the animal can escape to a cooler area of the cage if necessary (e.g., half on the pad and half off). If the logistics of using electric heating pads or circulating warm water blankets prevents their use for post-procedural heat support, air-activated thermal devices may provide a flexible, low-cost option (Beale et al. 2018). For procedures that involve prolonged anesthesia or a slow return to normal function, it may be desirable to increase the ambient temperature in the room for up to several days. If recovery is prolonged, reapplication of ophthalmic ointment will protect against desiccation of the corneas. Once the mouse is able to move around the cage, it should be checked frequently to make sure that it is eating and drinking. The provision of soft, moistened food will encourage eating if the mouse seems to be ignoring its normal hard food. Mice that are not feeling well will be more inclined to drink from a water bottle than from an automatic waterer. Animals that are slow to resume eating and drinking often are more willing to consume a highly palatable, nutrient-dense gel.

Analgesia

Administration of an analgesic is recommended for procedures that may involve post-procedural pain. Even in studies in which analgesia has historically been withheld because of concerns for interference with study objectives, emerging evidence suggests that analgesics can be successfully administered without significant alterations to study outcome (Bratcher et al. 2019; Herndon et al. 2016; Kennedy et al. 2014). Helpful reviews on the impact of pain and analgesia on various models are available (Carpenter et al. 2019; DeMarco and Nunamaker 2019; Huss et al. 2019; Larson et al. 2019; Taylor 2019; Toth 2019). When requests to withhold analgesics are submitted, IACUCs should carefully review the justification and recent literature and weigh the potential impact of withholding analgesia against the physiological and behavioral changes associated with pain that can also impact study results.

Signs of pain in mice include the following:

- Changes in facial expression to include orbital tightening, nose and/or cheek bulging, flattened ear position, and changes to whisker position, collectively known as the Mouse Grimace Scale (MGS) (Langford et al. 2010);
- Partial/complete closing of the eyelids; sunken eyes;
- Changes in respiration, which may include increased or decreased, shallow or labored respiratory patterns;
- Rough hair coat from lack of grooming; incontinence with soiled hair coat;
- Increased or decreased vibrissal movements;
- Decreased or hyperactive responsiveness to handling, or withdrawal from other mice in the group;
- Writhing, scratching, biting, or self-mutilation;

- Hunched posture;
- Sudden, sharp movement, such as running;
- Vocalization when being handled or palpated;
- Dehydration or weight loss, with wasting of the muscles on the back and a sunken or distended abdomen;
- Ataxia or circling;
- Hypothermia;
- Increased time or failure to build a complex nest post-procedurally (Gallo et al. 2020; Gaskill et al. 2013; Jirkof et al. 2013; Oliver et al. 2018; Rock et al. 2014).

Many drugs are available that provide effective analgesia in mice. The best strategy to minimize post-procedural pain with minimal stress to the animal involves pre-emptive (given in a time frame so that effective analgesic blood levels are reached before recovery from anesthesia) administration of an analgesic, minimization of stressors in the animal's environment, and regular postoperative assessment of the animal for indications that supplemental doses of analgesic are needed. Use of multiple analgesics with different mechanisms of action can provide an increased level of analgesia. For example, carprofen and buprenorphine might be given before surgery, followed by administration of topical bupivacaine after closure of the incision. Consult with your institution's veterinarian for recommendations on appropriate analgesic combinations based on the expected degree of post-procedural pain.

Consideration must be given to the route of administration of analgesics, to ensure that effective doses are provided. While multiple analgesics can be given in drinking water, this is a very inaccurate method for analgesic administration, and it is virtually impossible to ensure that mice are receiving adequate doses. Mice can be neophobic, and, if the first time they experience a new taste because of the addition of analgesic medication to their drinking water is when they are in pain, they may associate the new taste with pain and be even less likely to drink. Thus, in addition to lack of appropriate analgesia, mice will become dehydrated as well. Further, analgesics can precipitate, leading to possible overdoses as the medication settles to the bottom of the water bottle. Finally, particularly with group-housed mice, it is impossible to gauge how much water a given mouse is drinking, and therefore it is impossible to determine if appropriate analgesic doses are being ingested. For this reason, administration of analgesics in the drinking water is strongly discouraged. With the development of sustained release formulations of both NSAIDs and buprenorphine (see Table 6.2), several days of analgesia can be provided with a single administration, offering the same convenience and decreased handling as water bottle administration. If analgesics must be administered in the drinking water, strong consideration should be given to a single intra-operative dose of injectable analgesia to ensure that animals are not in pain when they recover from anesthesia, and reducing the chance that they associate the new taste of their water with pain. Pre-emptive administration of analgesics in water several days prior to the painful procedure can also help prevent post-procedural neophobia. In any case, when analgesics are administered in the drinking water, this route should be justified in the IACUC protocol, and details should be provided on how appropriate dosing and monitoring for breakthrough pain will be accomplished.

As pain is augmented by other stressors, postoperative pain can be reduced indirectly by keeping the mouse clean and dry, protecting it from extremes of temperature, and providing it with sufficient quantities of clean water and palatable, nutritious food. Soft, moistened food or a highly palatable, nutrient-dense moisture source in gel

TABLE 6.2: ANALGESICS FOR MICE[a]

Analgesic	Dosage	Route	Duration of Effect
Buprenorphine	0.1–0.5 mg/kg	SC	4–6 h
Buprenorphine SR™ [b]	0.5–1 mg/kg	SC	48–72 h
Buprenorphine (Ethiqa) XR™ [c]	3.25 mg/kg	SC	Up to 72 h
Carprofen	5 mg/kg	SC	12 h
	20 mg/kg	SC	24 h
Meloxicam	5–10 mg/kg	SC	8–12 h
Meloxicam SR™ [b,d]	4 mg/kg	SC	24–72 h

[a] Standard formulations of analgesics can be diluted to assist with administration volumes. Sustained release formulations should never be diluted to avoid interference with their depot properties. Follow manufacturers' and your institution's guidance for expiration dates of stock solutions following vial puncture, and of diluted substances.

[b] SR = sustained release. SR formulations should never be diluted to avoid interference with their depot properties.

[c] XR = extended release. FDA-indexed pharmaceutical-grade extended-release buprenorphine (Fidelis Pharmaceuticals 2020).

[d] Efficacy studies in mice have suggested that the duration of action is limited to 24 h or less (Kendall et al. 2014; Herrod et al. 2017). Until further studies are published, use of Meloxicam SR in mice should be accompanied by careful observation for signs of breakthrough pain beyond 24 h post-administration.

form may be more appealing than regular rodent chow. Water may be more readily taken from water bottles than from automatic watering devices. Although mice that are accustomed to group housing may be stressed by isolation, efforts must be made to protect vulnerable animals (e.g., those recovering from surgery) from aggression by cagemates. Changes in the general environment may be quite stressful. It is important to minimize changes in food (other than softening/moistening the food or offering a gel supplement), water, lighting, social grouping, and position within the room. A possible exception to this is room temperature. Mice that are experiencing a difficult or painful recovery from surgery will often do better in a warmer than normal (but not excessively hot) environment.

For pain that cannot be controlled by the methods described above, it will be necessary to administer supplemental doses of analgesic drugs postoperatively. Some of the drugs that have proven effective for reducing pain in mice are listed in Table 6.2. Note that, as we learn more about the efficacy of various analgesics in mice, dose recommendations change (Clark et al. 2014; Kendall et al. 2014, 2016; Matsumiya et al. 2012; Miller et al. 2016; Oliver et al. 2018; Sarfaty et al. 2019; Foley et al. 2019). Regular review of the literature, combined with careful post-procedural monitoring for breakthrough pain and consultation with veterinarians at your institution, is strongly recommended to ensure that adequate post-procedural analgesia is achieved.

euthanasia

No work on euthanasia escapes the definition of the term. The word *euthanasia* is derived from the Greek, meaning "easy or good death" (*Merriam-Webster*, 1995). In keeping with this definition, euthanasia must, above all, be as humane as possible. This means that, ideally, euthanasia causes a rapid loss of consciousness without pain or fear, followed by a reliable progression to death without regaining consciousness. To achieve these goals, consideration must be given to the method to be used, the time and place where the euthanasia will be performed, and appropriate training of the person who will perform the euthanasia. General recommendations are that animals are not euthanized near to other animals, although there is limited evidence that this disturbs mice (Boivin et al. 2016). Thought should also be given to the possible effects of the procedure on human participants or onlookers, possible effects on the scientific goals of the study, and practical constraints that would render one method more desirable than another.

Laws, Regulations, and Guidelines

Euthanasia of mice in conjunction with biomedical research, testing, or education in US institutions receiving National Institutes of Health (NIH) funding must be performed in accordance with the Public Health Service (PHS) Policy on Humane Care and Use of Laboratory Animals (Public Health Service, 2015) (Public Health Service 1986). PHS policy requires adherence to the provision of the Animal Welfare Act (Code of Federal Regulations, 1992) (n.d.), the *Guide for the Care and Use of Laboratory Animals* (National Research Council 2011), and the most recent report of the American Veterinary Medical Association (AVMA) Panel on Euthanasia (Leary et al. 2020). All of these documents emphasize the importance of: (1) performing euthanasia in a humane manner and (2) assuring that people charged with performing euthanasia have the necessary training and skills to perform it correctly. The report of the AVMA Panel on Euthanasia contains considerably more detail on the implications of euthanasia and the advantages and disadvantages of specific techniques that can be used. Some US states have specific guidelines regarding methods of euthanasia. It is important to know the regulations in your state. The institutional veterinarian can be consulted regarding the details of federal and state laws and guidelines pertaining to euthanasia.

Management Considerations

Euthanasia may affect people or other aspects of facility management in several ways. The necessity of assuring that personnel who will perform euthanasia are fully trained in the method(s) to be used has already been mentioned. Other points to consider include the following:

- The psychological implications of euthanasia. The act of taking life, by any method, can be difficult for humans performing the euthanasia, especially if it must be performed often or if an emotional attachment to the animal has developed. Resources on compassion fatigue are widely available.
- The aesthetic implications of different methods. Some methods (e.g., decapitation) can be unpleasant for human participants and observers.

- The health and safety of humans and other animals. Some methods of euthanasia are inherently more dangerous than others (e.g., decapitation by guillotine or carbon monoxide).
- The potential for human abuse. Some methods require the use of substances that may be abused by humans (e.g., barbiturates). Appropriate facilities and procedures must exist to control access to these substances. In the United States, controlled substances such as barbiturates may only be purchased by individuals or institutions registered with the Drug Enforcement Administration (DEA). Detailed records must be kept of the use and disposal of controlled substances.
- The expense and availability of drugs and equipment needed for some methods of euthanasia. In addition, the difficulty and expense of maintaining equipment in proper working order may be a consideration.

Scientific Considerations

It is important to recognize that the method of euthanasia affects postmortem evaluations or other scientific objectives of the research. Possible effects on scientific goals should be considered when choosing among the various methods that can be used to euthanize mice.

Experimental Endpoints

In studies where it is critical to scientific objectives to know when an animal would die, investigators should identify the earliest endpoints that predict an inevitable progression to death within a definable time frame. Properly conceived endpoints will limit animal suffering without compromising scientific objectives (Dunlap 2015; Workman et al. 2010). Endpoints that both reduce terminal distress for the animal and support scientific objectives must be determined with care based on knowledge of the specific model being studied. For some studies, appropriate endpoints can be identified based on previous experience with the model or published data from similar studies. However, for other studies, humane endpoints may not be known or predictable based on experience or data in the published literature. If no specific criteria are known to predict death within an acceptable (as defined by scientific objectives) time frame, investigators should attempt to identify such criteria for future studies. Pilot studies to establish humane endpoints may be appropriate in such cases. Identification of humane endpoints that predict an inevitable progression to death will require careful, and usually more frequent, observation of animals as their condition deteriorates and death approaches. More on this topic can be found in Chapter 4 of this handbook ("clinical medicine").

Euthanasia Methods

The following methods are acceptable for euthanasia of mice.

1. **Inhalant anesthetic overdose (isoflurane, sevoflurane, enflurane, or desflurane, with or without nitrous oxide)**: Effective scavenging of waste gases is essential to protect humans and other animals. This method is not appropriate for neonates, except with prolonged exposure and careful checking to assure death prior to disposal. Methoxyflurane and halothane are also acceptable but not widely available.

2. **Carbon dioxide**: To minimize pain, conscious animals should be exposed to chamber displacement rates of 30–70% chamber volume/min. It is important the chamber is not pre-filled with CO_2. This method is not appropriate for neonates unless exposure times are prolonged or an adjunct method is used to assure death (Pritchett et al. 2005). If an adjunct method is not used to assure death, animals must be examined carefully to assure that they are dead prior to disposal.

3. **Barbiturate overdose**: Pentobarbital (100 mg/kg, IP) is the most commonly used barbiturate for euthanasia of mice.

4. **Microwave irradiation**: Microwave ovens specifically designed for euthanasia are sometimes used for mice. An oven designed for food preparation must never be used for this purpose.

The following methods are acceptable for euthanasia of conscious, adult mice only if certain conditions are met.

1. **Carbon monoxide (CO)**: Appropriate precautions (good ventilation and monitors) must be taken to prevent human exposure. One should use compressed CO only. The concentration of CO in the euthanasia chamber should be at least 6%.

2. **Decapitation**: Scientific justification and IACUC approval are required, and appropriate training and demonstrated skill are essential. Performance of this technique requires either a small guillotine or heavy, sharp scissors. Decapitation is potentially dangerous to humans (guillotine), is aesthetically disagreeable to many individuals, and may cause significant pain to the mouse if performed incorrectly (i.e., with dull blades and/or anything less than a single swift, bold stroke or cut). The potential for pain to the mouse is reduced or eliminated if the mouse is first anesthetized. Many believe that, if done properly, this is the most humane method of euthanasia for neonatal mice.

3. **Cervical dislocation**: Scientific justification and IACUC approval are required, and appropriate training and demonstrated skill are absolutely essential. Cervical dislocation does not require specialized equipment, but mechanical cervical dislocators are commercially available and can be quite useful. Like decapitation, cervical dislocation is aesthetically disagreeable and may cause significant pain for the mouse if performed incorrectly. The potential for pain to the mouse is reduced or eliminated if the mouse is first anesthetized. Many believe that, if done correctly, cervical dislocation is a humane form of euthanasia for mice. Anyone wishing to use this technique should be thoroughly trained by someone who is proficient in its performance.

The following methods are unacceptable for euthanasia of conscious, adult mice.

1. **Potassium chloride**: When injected IV or directly into the heart of a *fully anesthetized* animal, potassium chloride produces immediate cardiac arrest and death. Potassium chloride must never be used to euthanize a conscious animal.

2. **Ether**: Because of safety concerns (ether is both flammable and explosive), this inhalant drug should be used only under carefully controlled conditions that meet all regulatory and safety guidelines.

3. **Nitrogen, argon**: The animal must be heavily sedated or anesthetized prior to placement in an air-tight container that has been prefilled with one of these gases, after which more nitrogen or argon is rapidly piped in. Death results quickly

from lack of oxygen (hypoxemia). As euthanasia with nitrogen or argon is distressful to mice, other methods of euthanasia are preferred.

---

necropsy

Necropsy (viewing the dead) refers to the postmortem examination of organs and tissues (this procedure in humans is called an autopsy). A useful necropsy technique should allow efficient evaluation of all organs and lesions and should be teachable and reproducible (Hampshire and Rippy 2015; Parkinson et al. 2011; Scudamore et al. 2014; Treuting and Snyder 2015). A detailed plan and checklist can help ensure that all procedures are performed (e.g., photography, radiography, and gross necropsy), and that all tissues are examined and weighed or measured. There are many ways to perform a necropsy. Variations in methods can be justified by primary aims to diagnose disease conditions or to achieve the specific aims of research studies. However, a standardized systematic procedure, to which minor alterations can be made, can improve comparisons within and between studies, even when initial examinations are months or years apart. Although not technically a necropsy, but rather a terminal surgery, research needs may require perfusion of the whole mouse with a fixative. This will not be discussed further, but online guides are available (Gage et al. 2012).

Strategic use of pathology in research settings can include clinical and anatomic pathology to:

1. Assess disease problems, including decreased production in breeding colonies;
2. Assess disease problems in research animals used in studies; and
3. Confirm, characterize, and validate research endpoints and phenotypes.

Equipment and Materials

1. **Workstation**: A ventilated workstation such as a down-draft table or a fume hood should be used to protect the prosector (person performing the necropsy) from fixative fumes or other hazardous materials.

2. **Eye protection**: Glasses or goggles should be worn to protect the eyes from splashes with fixative or other potentially
hazardous materials. The use of a dissecting microscope may
facilitate dissection and examination of small specimens.

3. **Face protection**: Face masks or respirators may be necessary to protect sensitive individuals from mouse allergens.

4. **Gloves**: When multiple animals will be examined, double
gloving may protect the hands better than frequent glove
changes. When the top gloves are damaged or soiled, they can
be replaced with minimal exposure of the hands to drying or
contaminating materials.

5. **Protective uniform**: Lab coats or other apparel should be
worn to protect skin and clothes from fixatives and contaminating materials.

6. **Cutting board**: An inexpensive plastic cutting board is adequate
for most purposes. It should be relatively easy to clean and able
to withstand frequent use. Some necropsy methods use pins
and soft or porous cutting boards. Disposable necropsy boards
with integrated measuring scales are available as well.

7. **Paper towels**: Many tissues (e.g., skin and reproductive tract)
can be laid flat on a paper towel or index card and will adhere
to it to facilitate examination and ensure uniform fixation.

8. **Scale or balance**: Animals should be weighed before dissection. When a body part is reported as being small or large, it
should be measured and/or weighed.

9. **Metric ruler**: When a finding or organ is reported as being
small or large, it should be measured and/or weighed. Masses
and organs with three dimensions should have measurements
recorded for three dimensions. For example, a spot may be 2 ×
3 mm, or a mass may be 2 × 3 × 2 mm. A ruler (or ruled label)
should be included in photographs to facilitate subsequent
measurements.

10. **Forceps**: Blunt-ended, serrated, or toothed forceps are desirable as they seem to cause the least damage to delicate mouse
tissues. Fine-pointed forceps can create artifactual holes or
tears in tissues. Smooth forceps require considerable compression to grip slippery tissue.

11. **Scissors**: Fine, blunt-ended scissors seem to cause the least
damage to delicate tissues. Sharp-tipped corneal scissors are

used commonly but can make holes or tears, especially in inexperienced hands. Two pair of scissors are recommended; one should be used only on bone and will need to be sharpened or replaced more frequently. Scissors with tungsten carbide blades may be useful for cutting bones.

12. **Scalpel blades**: These should be used sparingly in mouse dissection, but sharp, fresh blades are critical to trimming tissues for histology processing. Inexpensive single-edge blades may work as well or better than scalpel blades and handles.

13. **Syringe and needle**: A 3-ml syringe with 21-gauge needle works well for infusing the lungs and gastrointestinal tract with formalin.

14. **Fixative**: To preserve and prepare tissues for further processing, tissues are normally submerged in fixative solution. Neutral buffered formalin (10%) is suitable for soft tissues in most situations.

15. **Decalcifying (demineralizing) solution**: Bony tissues (e.g., head, spine, legs) should be decalcified for histology processing and evaluation. Some formic acid-based solutions fix tissues and demineralize simultaneously, so that mouse tissues are fully fixed and adequately demineralized for paraffin processing within 24 h. Once adequately demineralized, tissues should be processed promptly because overexposure to demineralizing agents (acids and or chelators) will damage tissues. Immunohistochemical techniques may or may not work on demineralized tissues.

16. **Camera**: Useful for documenting findings.

17. **Specimen containers** with labels.

18. **Labelled tissue cassettes**, if desired, to separate tissues.

Necropsy Procedure

Ideally, the animal should be evaluated immediately after death. Because of their small size, mice autolyze quickly, and so animals are ideally euthanized and then dissected, not necropsied after being found dead. Alternatively, carcasses may be stored for a short time (several hours) under refrigeration to delay tissue decomposition. If there is no time to perform an adequate necropsy, carcasses can be opened (abdomen, chest, skull) and put entire into a 480 ml 10%NBF container. The formalin will preserve the mouse, and the histology technician can retrieve the organs at the time of processing. Carcasses and any other research materials or specimens should be kept in designated refrigerators not used for storage of food for animals or personnel. Do not freeze carcasses; freezing compromises gross pathology and histopathology. A systematic, reproducible approach to necropsy should be developed using the general procedures described below.

External examination

The animal should be weighed and then examined to determine sex, color and general condition of the coat, eye color, and general body condition (e.g., thin, adequate or good body condition, or obese). External lesions (e.g., domed head, anophthalmia, masses, or open wounds) should be described and measured.

Gentle palpation may reveal pups or other abdominal masses or suggest the presence of fluid. A sterile sample can be obtained with a needle and syringe for chemistry (e.g., BUN, protein), cytology, or microbiology evaluation. The consistency of any palpated masses should be described as soft or fluctuant, firm or hard. Hard should be reserved for bony or mineralized masses.

Dissection and specimen collection

It is helpful to always orient animals in the same direction (e.g., head up or head right) so that the side of the lesion can be recalled accurately. It is useful to weigh organs and to record descriptions of any lesions, including color, size, and/or weight. Necropsy dissection methods vary with the training of the prosectors, resources at the site, and needs of the projects.

Specific steps for one dissection method follow.

1. Remove the pelt by incising the ventral abdominal skin and exerting gentle pressure cranially and caudally until the pelt has been removed. This facilitates assessment of subcutaneous fat (minimal, adequate, or abundant) and mammary tissue in female mice, and reveals subcutaneous lesions and abdominal organs in situ.

2. After examining the animal with the skin off, begin at the head and remove the "chain" of parotid, sublingual, and

submandibular salivary glands, and the lymph nodes that extend from ear to ear under the chin.

3. Open the abdomen, xiphoid to pubis, and examine the contents in situ.

4. Lift the sternum by the xiphoid process and cut the ribs on either side to remove it and expose the thoracic cavity. Examine the contents, noting fluid or masses, and absence or enlargement of the thymus.

5. Split the mandibular symphysis with scissors to expose the tongue. Grasp the tongue with forceps and gently retract caudally to remove it, along with the larynx, trachea, and esophagus. Continue retracting to remove the heart, thymus, and lungs, with aorta, esophagus, and trachea, from the thorax. Use blunt dissection with the scissors to free these tissues. Examine the oral cavity.

6. Use the 3-ml syringe and 21-gauge needle to infuse the lungs with fixative. The lungs should expand fully, and excess fixative will reflux up the trachea. Difficulty in infusion may be due to inflammatory or neoplastic process. It is not necessary to clamp or tie the trachea.

7. Examine the removed viscera (tongue to diaphragm). The thyroid glands are immediately caudal to the larynx on either side of the trachea. They may be difficult to see without magnification, but they usually are included and can be examined microscopically on cross section if a 2–4-mm section of trachea immediately caudal to the larynx is removed.

8. Split the pelvis along the pubic symphysis to facilitate complete removal of abdominal contents. Insert closed scissors into the pelvic canal and open them gently to separate the halves of the pelvis.

9. Remove and examine the abdominal contents. Grasp the diaphragm with the forceps, cut at its deepest extent, and retract gently to lift out all of the abdominal contents together. The adrenal glands and kidneys tend to remain, deep in the retroperitoneal space; blunt dissection is usually required for their removal. Abdominal contents can be examined individually, organs weighed, and any abnormalities recorded.

10. Remove the liver and spleen for examination. The liver's small caudate lobe may be folded into the lesser curvature

of the stomach, and care should be taken to ensure that it is removed so that the entire liver is included when weighing that organ. When manipulating the liver, lift it gently or grasp parts that will not be submitted for histology (e.g., diaphragm or smaller liver lobes). The median and left lateral lobes are the largest and are usually selected for histology, unless lesions are present in other lobes or if experimental protocol dictates otherwise. The gallbladder lies between the two "halves" of the median lobe.

11. Remove the kidneys and adrenals for examination. The right kidney and adrenal should be anterior to the left. Female adrenal glands normally are larger than male adrenal glands. If it is important to distinguish right from left kidney, one kidney may be incised, and that incision recorded. A common mnemonic is "left, longitudinal," indicating that the left kidney is the one with the longitudinal incision.

12. Remove the reproductive tract. Tissue caudal to the kidneys that is not gastrointestinal tract is mostly reproductive tract and fat. Laying it flat on a dry piece of paper and spreading the tissue into its anatomic orientation can facilitate examination, fixation, and histology. The paper with attached tissue can be submerged in fixative.

13. Infuse and extend the gastrointestinal tract. Infuse different segments with 0.5–1.0 ml of formalin fixative using the 3-ml syringe and 21-gauge needle. Extend the GI tract gently by grasping the stomach in one hand and the rectum (fecal balls) in the other and separate gently to break the mesenteric attachments. Remove the attached mesentery, lymph nodes, and pancreas. These pale, soft tissues may be difficult to distinguish grossly but can be placed into a cassette and fixed for histology examination. Making a Swiss roll of the intestinal tract is also possible (Moolenbeek and Ruitenberg 1981), and some pathologists prefer this technique.

14. Submerge all tissues to be saved in at least ten times their volume of fixative.

15. Bone must be decalcified (demineralized) for routine histology processing in paraffin. Skin and other soft tissues should be removed so that the solution can penetrate the bone. If marrow from the sternum will be evaluated, the sternum should not be decalcified. Formic acid–based fixing and decalcifying

solutions can demineralize most mouse tissue satisfactorily within 24 h. Overexposure to the acid will digest the tissue and compromise evaluation. The ratio of tissue to solution should be approximately 1:10, and the tissue should be completely covered by the solution.

16. Photography: Digital images become part of the data and can be important for publication of findings. Pathology photographs should illustrate the lesion without horrifying viewers. Focus on the lesion, not the parts of the animal that make it "cute" (no faces or feet). Orientation of the animal or specimens should be consistent, and attention should be paid to identification or labeling of images.

Trimming tissues for histology

Consistent reproducible tissue trimming for histology processing improves the resulting slides and histopathology and facilitates comparisons within and between studies. Published recommendations for trimming for regulatory-type toxicity studies offer useful and very detailed guidance (Golubeva and Rogers 2009; Kittel et al. 2004; Morawietz et al. 2004; Ruehl-Fehlert et al. 2003). If tissues are to be used for molecular study, further care should be taken. The Appendix includes trimming suggestions and a numbering system for evaluating more than 40 mouse tissues on ten slides.

During dissection, some tissues can be placed directly in histology cassettes and submitted for histology processing as is. These can include lungs, trachea with thyroid, salivary glands, lymph nodes, pancreas, and reproductive tract when it is small. Other tissues must be "trimmed" further before being placed in cassettes. Trimming should be performed in a well-ventilated area or hood. Used fixative should be discarded as hazardous waste. After trimming, labeled cassettes should be placed in clean formalin. Fixed tissues to be saved should be stored in sufficient clean fixative to keep them moist.

During trimming, tissues should be cut with a single clean swipe of a sharp blade, not squashed or sawed. Inexpensive single-edge blades are suitable for trimming most tissues, and they should be replaced as soon as they become dull so that tissue is not damaged. Trimmed specimens should be less than 3 mm deep to fit in standard cassettes without generating grid marks and "squish artifact." For decalcified skulls, trimming may be facilitated by use of Weck blades, which have a longer, sharper cutting edge. Decalcified tissue should be soft and easily cut. Crunchy tissue requires additional decalcification. Decalcified specimens should be rinsed in water, and cassettes can be kept for short periods in buffered saline until histology processing.

Numbering of cassettes should be systematic to facilitate retrieval of specific tissues from archived material. A regular graphite pencil is the usual marker of choice because many inks are removed by alcohols during paraffin processing.

Histopathology

Diagnostic criteria and terminology for gross findings and for histopathology findings should be standardized to facilitate comparisons within and between studies. Anatomy and pathology terminology should be consistent with widely used and published systems to facilitate communication and publication (Hampshire and Rippy 2015; Parkinson et al. 2011; Treuting and Snyder 2015).

Reporting and archiving data and specimens

Data handling and reporting strategies vary with the resources and goals of the program or project. Paper copies of reports with a checklist for each body system or tissue may be sufficient for some projects. Databases, servers, and specialized hardware and software may be essential to the success of large projects and multidisciplinary and collaborative efforts. Archiving and reporting systems should be appropriate to preserve and protect data and specimens, to make them accessible for further evaluation, and to facilitate comparisons within and between studies. Pathology data frequently include many large images that require significant data handling capabilities and server space. Still images can exceed 5 MB, and virtual microscopy images can exceed 400 MB. Archiving of specimens, including fluids, wet tissues, frozen tissues, paraffin blocks, glass slides, and virtual (digital) slides, can require significant planning regarding preservation methods, identification, storage methods and capacity, and retrieval.

references

Beale CN, Esmail MY, Aguiar AM, Coughlin L, Merley AL, Alarcon Falconi TM, Perkins SE. 2018. Use of air-activated thermal devices during recovery after surgery in mice. *J Am Assoc Lab Anim Sci* **57**:392–400.

Boivin GP, Bottomley MA, Grobe N. 2016. Responses of male C57BL/6N mice to observing the euthanasia of other mice. *J Am Assoc Lab Anim Sci* **55**:406–411.

Braden GC, Brice AK, Hankenson FC. 2015. Adverse effects of vapocoolant and topical anesthesia for tail biopsy of preweanling mice. *J Am Assoc Lab Anim Sci* **54**:291–298.

Bratcher NA, Frost DJ, Hickson J, Huang X, Medina LM, Oleksijew A, Ferguson DC, Bolin S. 2019. Effects of buprenorphine in a preclinical orthotopic tumor model of ovarian carcinoma in female CB17 SCID mice. *J Am Assoc Lab Anim Sci* **58**:583–588.

Broome RL, Feng L, Zhou Q, Smith A, Hahn N, Matsui SM, Omary MB. 1999. Non-invasive transgenic mouse genotyping using stool analysis. *FEBS Lett* **462**:159–160.

Byers SL, Wiles MV, Dunn SL, Taft RA. 2012. Mouse estrous cycle identification tool and images. *PloS One* **7**:e35538.

Carpenter KC, Hakenjos JM, Fry CD, Nemzek JA. 2019. The influence of pain and analgesia in rodent models of sepsis. *Comp Med* **69**:546–554.

Chatelain G, Brun G, Michel D. 1995. Screening of homozygous transgenic mice by comparative PCR. *BioTechniques* **18**:958–960, 962.

Clark TS, Clark DD, Hoyt RF, Jr. 2014. Pharmacokinetic comparison of sustained-release and standard buprenorphine in mice. *J Am Assoc Lab Anim Sci* **53**:387–391.

Code of Federal Regulations. Title 9, Chapter 1, Subchapter a: animal welfare: part 2 regulations (§2.31).

Constantinescu GM, Duffee NE. 2017. Comparison of submental blood collection with the retroorbital and submandibular methods in mice (*Mus musculus*). *J Am Assoc Lab Anim Sci* **56**:711–712.

Danneman PJ, Mandrell TD. 1997. Evaluation of five agents/methods for anesthesia of neonatal rats. *Lab Anim Sci* **47**:386–395.

DeMarco GJ, Nunamaker EA. 2019. A review of the effects of pain and analgesia on immune system function and inflammation: relevance for preclinical studies. *Comp Med* **69**:520–534.

Desjardins C. 1986. Indwelling vascular cannulas for remote blood sampling, infusion, and long-term instrumentation of small laboratory animals, pp. 143–194. In *Research Surgery and Care of the Research Animal Part A Patient Care, Vascular Access, and Telemetry*, Gay WI, Heavner, J, Eds. Orlando: Academic Press, Inc.

Doerning CM, Thurston SE, Villano JS, Kaska CL, Vozheiko TD, Soleimanpour SA, Lofgren JL. 2019. Assessment of mouse handling techniques during cage changing. *J Am Assoc Lab Anim Sci* **58**:767–773.

Dudley ES, Johnson RA, French DC, Boivin GP. 2016. Effects of topical anesthetics on behavior, plasma corticosterone, and blood glucose levels after tail biopsy of C57BL/6NHSD mice (*Mus musculus*). *J Am Assoc Lab Anim Sci* **55**:443–450.

Dunlap J. 2015. Humane endpoints for animals used in training. *Lab Anim* **44**(2):71.

Erickson RL, Terzi MC, Jaber SM, Hankenson FC, McKinstry-Wu A, Kelz MB, Marx JO. 2016. Intraperitoneal continuous-rate infusion for the maintenance of anesthesia in laboratory mice (Mus *musculus*). *J Am Assoc Lab Anim Sci* **55**:548–557.

Fidelis Pharmaceuticals. 2020. *Take the Pain out of Pain Management.* Available at: https://ethiqaxr.com/wp-content/uploads/2020/08/fid-eth-001_sales_aid_mech_v5_low_rez_single_pages.pdf.

Foley PL, Kendall LV, Turner PV. 2019. Clinical management of pain in rodents. *Comp Med* **69**:468–489.

Frohlich JR, Alarcón CN, Toarmino CR, Sunseri AK, Hockman TM. 2018. Comparison of serial blood collection by facial vein and retrobulbar methods in C57BL/6 mice. *J Am Assoc Lab Anim Sci* **57**:382–391.

Gaertner DJ, Hallman TM, Hankenson FC, Batchelder MA. 2008. Anesthesia and analgesia for laboratory rodents, pp. 239–297. In *Anesthesia and Analgesia in Laboratory Rodents*, Fish R, Danneman P, Brown M, Karas A, Eds., 2nd ed. Boston: Academic Press.

Gage GJ, Kipke DR, Shain W. 2012. Whole animal perfusion fixation for rodents. *J Vis Exp* **65**:e3564. https://doi.org/10.3791/3564.

Gallo MS, Karas AZ, Pritchett-Corning K, Garner Guy Mulder JP, Gaskill BN. 2020. Tell-tale TINT: does the time to incorporate into nest test evaluate postsurgical pain or welfare in mice? *J Am Assoc Lab Anim Sci* **59**:37–45.

Garzel LM, Hankenson FC, Combs J, Hankenson KD. 2010. Use of quantitative polymerase chain reaction analysis to compare quantity and stability of isolated murine DNA. *Lab Anim* **39**:283–289.

Gaskill BN, Karas AZ, Garner JP, Pritchett-Corning KR. 2013. Nest building as an indicator of health and welfare in laboratory mice. *J Vis Exp* **82**:51012. https://doi.org/10.3791/51012

Gehling AM, Kuszpit K, Bailey EJ, Allen-Worthington KH, Fetterer DP, Rico PJ, Bocan TM, Hofer CC. 2018. Evaluation of volume of intramuscular injection into the caudal thigh muscles of female and male BALB/c mice (*Mus musculus*). *J Am Assoc Lab Anim Sci* **57**:35–43.

Golubeva Y, Rogers K. 2009. Collection and preparation of rodent tissue samples for histopathological and molecular studies in carcinogenesis. *Methods Mol Biol* **511**:3–60.

Gouveia K, Hurst JL. 2019. Improving the practicality of using non-aversive handling methods to reduce background stress and anxiety in laboratory mice. *Sci Rep* **9**:20305.

Hamacher J, Arras M, Bootz F, Weiss M, Schramm R, Moehrlen U. 2008. Microscopic wire guide-based orotracheal mouse intubation: description, evaluation and comparison with transillumination. *Lab Anim* **42**:222–230.

Hampshire V, Rippy M. 2015. Optimizing research animal necropsy and histology practices. *Lab Anim* **44**:170–172.

Hankenson FC, Garzel LM, Fischer DD, Nolan B, Hankenson KD. 2008. Evaluation of tail biopsy collection in laboratory mice (*Mus musculus*): vertebral ossification, DNA quantity, and acute behavioral responses. *J Am Assoc Lab Anim Sci* **47**:10–18.

Hayward AM, Lemke LB, Bridgeford EC, Theve EJ, Jackson CN, Cunliffe-Beamer TL, Marini RP. 2007. Biomethodology and surgical techniques, pp. 437–488. In *The Mouse in Biomedical Research*, vol **III** (Normative Biology, Husbandry, and Models), Fox JG, Barthold SW, Davisson MT, Newcomer CE, Quimby FW, Smith AL, Eds. Boston: Academic Press, Inc.

Heimann M, Kasermann HP, Pfister R, Roth DR, Burki K. 2009. Blood collection from the sublingual vein in mice and hamsters: a suitable alternative to retrobulbar technique that provides large volumes and minimizes tissue damage. *Lab Anim* **43**:255–260.

Heimann M, Roth DR, Ledieu D, Pfister R, Classen W. 2010. Sublingual and submandibular blood collection in mice: a comparison of effects on body weight, food consumption and tissue damage. *Lab Anim* **44**:352–358.

Herndon NL, Bandyopadhyay S, Hod EA, Prestia KA. 2016. Sustained-release buprenorphine improves postsurgical clinical condition but does not alter survival or cytokine levels in a murine model of polymicrobial sepsis. *Comp Med* **66**:455–462.

Hoffman JF, Fan AX, Neuendorf EH, Vergara VB, Kalinich JF. 2018. Hydrophobic sand versus metabolic cages: a comparison of urine collection methods for rats (*Rattus norvegicus*). *J Am Assoc Lab Anim Sci* **57**:51–57.

Hoffman JF, Vechetti IJ, Jr., Alimov AP, Kalinich JF, McCarthy JJ, Peterson CA. 2019. Hydrophobic sand is a viable method of urine collection from the rat for extracellular vesicle biomarker analysis. *Mol Genet Metab Rep* **21**:100505.

Hoffman JF, Vergara VB, Mog SR, Kalinich JF. 2017. Hydrophobic sand is a non-toxic method of urine collection, appropriate for urinary metal analysis in the rat. *Toxics* **5**(4):25. https://doi.org /10.3390/toxics5040025.

Huss MK, Felt SA, Pacharinsak C. 2019. Influence of pain and analgesia on orthopedic and wound-healing models in rats and mice. *Comp Med* **69**:535–545.

Irwin MH, Moffatt RJ, Pinkert CA. 1996. Identification of transgenic mice by PCR analysis of saliva. *Nat Biotechnol* **14**:1146–1148.

Janssen CF, Maiello P, Wright MJ, Jr., Kracinovsky KB, Newsome JT. 2017. Comparison of atipamezole with yohimbine for antagonism of xylazine in mice anesthetized with ketamine and xylazine. *J Am Assoc Lab Anim Sci* **56**:142–147.

Jirkof P, Fleischmann T, Cesarovic N, Rettich A, Vogel J, Arras M. 2013. Assessment of postsurgical distress and pain in laboratory mice by nest complexity scoring. *Lab Anim* **47**:153–161.

Kalippke K, Werwitzke S, von Hornung M, Mischke R, Ganser A, Tiede A. 2009. DNA analysis from stool samples: a fast and reliable method avoiding invasive sampling methods in mouse models of bleeding disorders. *Lab Anim* **43**:390–393.

Kendall LV, Hansen RJ, Dorsey K, Kang S, Lunghofer PJ, Gustafson DL. 2014. Pharmacokinetics of sustained-release analgesics in mice. *J Am Assoc Lab Anim Sci* **53**:478–484.

Kendall LV, Wegenast DJ, Smith BJ, Dorsey KM, Kang S, Lee NY, Hess AM. 2016. Efficacy of sustained-release buprenorphine in an experimental laparotomy model in female mice. *J Am Assoc Lab Anim Sci* **55**:66–73.

Kennedy LH, Hwang H, Wolfe AM, Hauptman J, Nemzek-Hamlin JA. 2014. Effects of buprenorphine and estrous cycle in a murine model of cecal ligation and puncture. *Comp Med* **64**:270–282.

Kerst J. 2003. An ergonomics process for the care and use of research animals. *ILAR J* **44**:3–12.

Kittel B, Ruehl-Fehlert C, Morawietz G, Klapwijk J, Elwell MR, Lenz B, O'Sullivan MG, Roth DR, Wadsworth PF. 2004. Revised guides for organ sampling and trimming in rats and mice--Part 2. A joint publication of the RITA and NACAD groups. *Exp Toxicol Pathol* **55**:413–431.

Lahm H, Hoeflich A, Rieger N, Wanke R, Wolf E. 1998. Identification of transgenic mice by direct PCR analysis of lysates of epithelial cells obtained from the inner surface of the rectum. *Transgenic Res* **7**:131–134.

Lamanna OK, Hsieh MH, Forster CS. 2020. Novel catheter design enables transurethral catheterization of male mice. *Am J Physiol Renal Physiol* **319**:F29–F32.

Langford DJ, Bailey AL, Chanda ML, Clarke SE, Drummond TE, Echols S, Glick S, Ingrao J, Klassen-Ross T, Lacroix-Fralish ML, Matsumiya L, Sorge RE, Sotocinal SG, Tabaka JM, Wong D, van den Maagdenberg AM, Ferrari MD, Craig KD, Mogil JS. 2010. Coding of facial expressions of pain in the laboratory mouse. *Nat Methods* **7**:447–449.

Larson CM, Wilcox GL, Fairbanks CA. 2019. Defining and managing pain in stroke and traumatic brain injury research. *Comp Med* **69**:510–519.

Leary SUW, Anthony R, Cartner S, Corey D, Grandin TGC, Gwaltney-Bran S, McCrackin MA, Meyer R, Miller DS, Yanong R. 2020. AVMA guidelines for the euthanasia of animals. [Cited 18 August 2021. Available at: https://www.avma.org/sites/default/files/2020-01/2020-Euthanasia-Final-1-17-20.pdf.

Levin-Arama M, Abraham L, Waner T, Harmelin A, Steinberg DM, Lahav T, Harley M. 2016. Subcutaneous compared with intra-peritoneal ketaminexylazine for anesthesia of mice. *J Am Assoc Lab Anim Sci* **55**:794–800.

Marshall BE, Longnecker DE. 1996. General anesthetics, pp. 307–330. In *Goodman & Gilman's The Pharmacological Basis of Therapeutics*, Hardman JG, Limbird LE, Molinoff PB, Ruddon RW, Gilman, AG, Eds. New York: McGraw-Hill.

Marx JO, Jensen JA, Seelye S, Walton RM, Hankenson FC. 2015. The effects of acute blood loss for diagnostic bloodwork and fluid replacement in clinically ill mice. *Comp Med* **65**:202–216.

Matsumiya LC, Sorge RE, Sotocinal SG, Tabaka JM, Wieskopf JS, Zaloum A, King OD, Mogil JS. 2012. Using the mouse grimace scale to reevaluate the efficacy of postoperative analgesics in laboratory mice. *J Am Assoc Lab Anim Sci* **51**:42–49.

Meldgaard M, Bollen PJ, Finsen B. 2004. Non-invasive method for sampling and extraction of mouse DNA for PCR. *Lab Anim* **38**:413–417.

Herrod JA, Doane CJ, Veltri CA, and Mexas AM. 2017. Inappropriate post-operative analgesia is achieved using recommended doses of sustained-release meloxicam in mice. *J Anim Health Behav Sci* **1**:109.

Meyer RE, Fish RE. 2008. Pharmacology of injectable anesthetics, sedatives, and tranquilizers. In *Anesthesia and Analgesia in Laboratory Rodents*, 2nd ed., Fish RE, Brown MJ, Danneman PJ, and Karas, AZ, Eds. Boston: Academic Press, 27.

Miller AL, Kitson GL, Skalkoyannis B, Flecknell PA, Leach MC. 2016. Using the mouse grimace scale and behaviour to assess pain in CBA mice following vasectomy. *Appl Anim Behav Sci* **181**:160–165.

Moolenbeek C, Ruitenberg EJ. 1981. The Swiss roll: a simple technique for histological studies of the rodent intestine. *Lab Anim* **15**:57–59.

Moore ES, Cleland TA, Williams WO, Peterson CM, Singh B, Southard TL, Pasch B, Labitt RN, Daugherity EK. 2017. Comparing phlebotomy by tail tip amputation, facial vein puncture, and tail vein incision in C57BL/6 mice by using physiologic and behavioral metrics of pain and distress. *J Am Assoc Lab Anim Sci* **56**:307–317.

Morawietz G, Ruehl-Fehlert C, Kittel B, Bube A, Keane K, Halm S, Heuser A, Hellmann J. 2004. Revised guides for organ sampling and trimming in rats and mice--Part 3. A joint publication of the RITA and NACAD groups. *Exp Toxicol Pathol* **55**:433–449.

National Research Council. 2011. *Guide for the Care and Use of Laboratory Animals*. Washington, DC: National Academies Press.

NC3Rs. *Mouse Handling FAQs*. [Cited 9/15/2020. Available at: https://www.nc3rs.org.uk/mouse-handling-faqs.

Oliver VL, Thurston SE, Lofgren JL. 2018. Using cageside measures to evaluate analgesic efficacy in mice (*Mus musculus*) after surgery. *J Am Assoc Lab Anim Sci* **57**:186–201.

Papaioannou VE, Fox JG. 1993. Efficacy of tribromoethanol anesthesia in mice. *Lab Anim Sci* **43**:189–192.

Parasuraman S, Raveendran R, Kesavan R. 2010. Blood sample collection in small laboratory animals. *J Pharmacol Pharmacother* **1**:87–93.

Parkinson CM, O'Brien A, Albers TM, Simon MA, Clifford CB, Pritchett-Corning KR. 2011. Diagnostic necropsy and selected tissue and sample collection in rats and mice. *J Vis Exp* **54**:e2966. https://doi.org/10.3791/2966.

Pritchett K, Corrow D, Stockwell J, Smith A. 2005. Euthanasia of neonatal mice with carbon dioxide. *Comp Med* **55**:275–281.

Public Health Service. 1986. *Policy on Humane Care and Use of Laboratory Animals*. Washington, DC: US Department of Health and Human Services; amended 2015.

Regan RD, Fenyk-Melody JE, Tran SM, Chen G, Stocking KL. 2016. Comparison of submental blood collection with the retroorbital and submandibular methods in mice (*Mus musculus*). *J Am Assoc Lab Anim Sci* **55**:570–576.

Reis LO, Sopena JM, Fávaro WJ, Martin MC, Simão AF, Reis RB, Andrade MF, Domenech JD, Cardo CC. 2011. Anatomical features of the urethra and urinary bladder catheterization in female mice and rats. An essential translational tool. *Acta Cir Bras* **26** Suppl 2:106–110.

Ren S, Li M, Cai H, Hudgins S, Furth PA. 2001. A simplified method to prepare PCR template DNA for screening of transgenic and knockout mice. *Contemp Top Lab Anim Sci* **40**:27–30.

Rock ML, Karas AZ, Rodriguez KB, Gallo MS, Pritchett-Corning K, Karas RH, Aronovitz M, Gaskill BN. 2014. The time-to-integrate-to-nest test as an indicator of wellbeing in laboratory mice. *J Am Assoc Lab Anim Sci* **53**:24–28.

Ruehl-Fehlert C, Kittel B, Morawietz G, Deslex P, Keenan C, Mahrt CR, Nolte T, Robinson M, Stuart BP, Deschl U. 2003. Revised guides for organ sampling and trimming in rats and mice--part 1. *Exp Toxicol Pathol* **55**:91–106.

Sarfaty AE, Zeiss CJ, Willis AD, Harris JM, Smith PC. 2019. Concentration-dependent toxicity after subcutaneous administration of meloxicam to C57BL/6N mice (*Mus musculus*). *J Am Assoc Lab Anim Sci* **58**:802–809.

Sato M, Kasai K, Tada N. 1995. A sensitive method of testing for transgenic mice using polymerase chain reaction-southern hybridization. *Genet Anal: Biomol Eng* **12**:109–111.

Schmitteckert EM, Prokop CM, Hedrich HJ. 1999. DNA detection in hair of transgenic mice--a simple technique minimizing the distress on the animals. *Lab Anim* **33**:385–389.

Scudamore CL, Busk N, Vowell K. 2014. A simplified necropsy technique for mice: making the most of unscheduled deaths. *Lab Anim* **48**:342–344.

Spoelstra EN, Ince C, Koeman A, Emons VM, Brouwer LA, van Luyn MJ, Westerink BH, Remie R. 2007. A novel and simple method for endotracheal intubation of mice. *Lab Anim* **41**:128–135.

St Clair MB, Sowers AL, Davis JA, Rhodes LL. 1999. Urinary bladder catheterization of female mice and rats. *Contemp Top Lab Anim Sci* **38**:78–79.

Taylor DK. 2019. Influence of pain and analgesia on cancer research studies. *Comp Med* **69**:501–509.

Toth LA. 2019. Interacting influences of sleep, pain, and analgesic medications on sleep studies in rodents. *Comp Med* **69**:571–578.

Treuting PM, Snyder JM. 2015. Mouse necropsy. *Curr Protoc Mouse Biol* **5**:223–233.

Vogler GA. 2008. Anesthesia delivery systems. In *Anesthesia and Analgesia in Laboratory Rodents*, 2nd ed,, Fish RE, Brown MJ, Danneman PJ, Karas AZ, Eds. Boston: Academic Press, 127.

Watanabe A, Hashimoto Y, Ochiai E, Sato A, Kamei K. 2009. A simple method for confirming correct endotracheal intubation in mice. *Lab Anim* **43**:399–401.

Winberg G. 1991. A rapid method for preparing DNA from blood, suited for PCR screening of transgenes in mice. *PCR Methods Appl* **1**:72–74.

Workman P, Aboagye EO, Balkwill F, Balmain A, Bruder G, Chaplin DJ, Double JA, Everitt J, Farningham DA, Glennie MJ, Kelland LR, Robinson V, Stratford IJ, Tozer GM, Watson S, Wedge SR, Eccles SA. 2010. Guidelines for the welfare and use of animals in cancer research. *Br J Cancer* **102**:1555–1577.

Yardeni T, Eckhaus M, Morris HD, Huizing M, Hoogstraten-Miller S. 2011. Retro-orbital injections in mice. *Lab Anim* **40**:155–160.

resources and additional information

Because this volume is intended to be a handbook, coverage is not exhaustive for most topics. In this regard, additional resources for information related to the care and use of laboratory mice are provided here.

organizations

A number of professional organizations exist that can serve as initial contacts for obtaining information regarding specific issues related to the care and use of laboratory mice. Membership of these organizations should be considered, since it allows the laboratory animal science professional to stay abreast of regulatory issues, improved procedures for the use of animals, management issues, and animal health issues. Relevant organizations include the following.

American Association for Laboratory Animal Science (AALAS), www.aalas.org

AALAS serves a diverse professional group, ranging from principal investigators to animal care technicians to veterinarians. The journals *Comparative Medicine* and *Journal of the American Association for Laboratory Animal Science* (JAALAS) are both published by AALAS and serve to communicate relevant information. *Laboratory Animal Science Professional* (LAS Pro) is a bimonthly magazine published by

DOI: 10.1201/9780429353086-7

AALAS that contains practical information on management, professional development, occupational health and safety, and technologies associated with laboratory animal science. AALAS sponsors a program for certification of laboratory animal science professionals at three levels: assistant laboratory animal technician (ALAT), laboratory animal technician (LAT), and laboratory animal technologist (LATG). Further, a certification program for managers of animal resource programs (CMAR) has been developed. An extensive online resource, the AALAS Learning Library, offers subscribers an extensive menu of courses relevant to laboratory animal science, including courses on biosafety. AALAS also sponsors an annual meeting and several discussion forums available through the AALAS Community Exchange (ACE). These include the Animal Researcher Community (ARC); CompMed, for professionals working in comparative medicine and biomedical research (including IACUC members and staff, and animal technicians); and an Open Forum for all of the AALAS membership. Local groups have also organized into smaller branches.

Laboratory Animal Management Association (LAMA), www.lama-online.org

LAMA serves as a mechanism for information exchange between individuals charged with management responsibilities for laboratory animal facilities. In this regard, the association publishes the *LAMA Review* and sponsors an annual meeting.

American Society of Laboratory Animal Practitioners (ASLAP), www.aslap.org

ASLAP is an association of veterinarians engaged in laboratory animal medicine. The society publishes a newsletter to foster communication between members and sponsors sessions at the annual AALAS meeting and the annual meeting of the American Veterinary Medical Association. The organization also represents laboratory animal practitioners through its seat in the AVMA House of Delegates.

American College of Laboratory Animal Medicine (ACLAM), www.aclam.org

ACLAM is an association of laboratory animal veterinarians founded to encourage education, training, and research in laboratory animal medicine. The ACLAM board certifies veterinarians in the specialty of laboratory animal medicine. The group sponsors the annual ACLAM Forum continuing education meeting, along with sessions at the annual AALAS and AVMA meetings.

Laboratory Animal Welfare Training Exchange (LAWTE), www.lawte.org

LAWTE is an organization of people who train in and for the laboratory animal science field. By sharing ideas on methods and materials for training, members can learn together how best to meet the training and qualification requirements of national regulations and guidelines. LAWTE holds a conference every 2 years for trainers to exchange information on their training programs in the United States and abroad.

Institute for Laboratory Animal Research (ILAR), www.nationalacademies.org/ilar/institute-for-laboratory-animal-research

The mission of ILAR is to evaluate and report on the scientific, technological, and ethical use of animals and related biological resources and of non-animal alternatives in non-food settings, such as research, testing, education, and production of pharmaceuticals. Using the principles of refinement, reduction, and replacement (3Rs) as a foundation, ILAR promotes high-quality science through the humane care and use of animals, and the implementation of alternatives. Through the reports of expert committees, the *ILAR Journal*, web-based resources, and other means of communication, ILAR functions as a component of the National Academies to provide independent, objective advice to the federal government, the international biomedical research community, and the public.

AAALAC International, www.aaalac.org

AAALAC International is a private, nonprofit organization that promotes the humane treatment of animals in science through voluntary accreditation and assessment programs. AAALAC provides a mechanism for peer evaluation of laboratory animal care and use programs. Accreditation by AAALAC is widely accepted as strong evidence of a quality research animal care and use program.

Foundation for Biomedical Research (FBR), fbresearch.org

FBR is a nonprofit organization dedicated to improving human and animal health by promoting public understanding and support for biomedical research. FBR, in collaboration with the National Association for Biomedical Research (NABR), creates educational resources such as webinars, brochures, posters, and white papers to benefit the public, the government, and research professionals.

National Association for Biomedical Research (NABR), www.nabr.org

NABR is a national nonprofit organization dedicated to advocating sound public policy for ethical and essential animal research. NABR is the leading public policy organization in support of animal research and represents the scientific community in federal, state, and local legislatures and at all levels of government. NABR works to educate legislators about the impact of proposed changes on medical advancements, and continues to play a central role in ensuring that new federal laws and regulations meet both animal welfare and biomedical research needs.

publications

A number of published materials are valuable as additional reference sources, including both books and periodicals.

Books

The following books may be worthwhile sources of additional information.

- *Laboratory Animal Medicine*, 3rd edition, edited by J.G. Fox, L.C. Anderson, G. Otto, K. R. Pritchett-Corning, and M.T. Whary, 2015, Academic Press.
- *The Mouse in Biomedical Research* (Volumes I–IV), edited by J.G. Fox, S.W. Barthold, M.T. Davisson, C.E. Newcomer, F.W. Quimby, and A.L. Smith, 2007, Academic Press.

- *A Colour Atlas of Anatomy of Small Laboratory Animals, Volume II: Rat, Mouse, Hamster,* by P. Popesko, V. Rajtová, and J. Horák, 1992, Wolfe, reprinted in 2002 by Saunders/Elsevier Science.
- *Pathology of Laboratory Rodents and Rabbits,* 4th edition, by S.W. Barthold, S.M. Griffey, and D. H. Percy, 2016, John Wiley.
- *Pathology of the Mouse,* edited by R.R. Maronpot, G.A. Boorman, and B.W. Gaul, 1999, Cache River Press.
- *Comparative Anatomy and Histology: A Mouse, Rat, and Human Atlas,* edited by P. Treuting, S. Dintzis, and K.S. Montine, 2017, Academic Press.
- *International Classification of Rodent Tumors: The Mouse,* edited by U. Mohr, 2001, Springer-Verlag.
- *Atlas of Mouse Hematopathology,* by T.N. Fredrickson and A.W. Harris, 2000, Harwood Academic.
- *The Anatomical Basis of Mouse Development,* by M.H. Kaufman and J.B.L. Bard, 1999, Academic Press.
- *Formulary for Laboratory Animals,* 3rd edition, by C.T. Hawk, S. Leary, and T. Morris, 2005, Wiley-Blackwell.
- *Clinical Laboratory Animal Medicine: An Introduction,* 5th edition, by L.A. Colby, M.H. Nowland, and L.H. Kennedy, 2020, John Wiley.
- *Mouse Hematology: A Laboratory Manual,* by M.P. McGarry, C.A. Protheroe, and J.J. Lee, 2010, Cold Spring Harbor Laboratory Press.
- *The IACUC Handbook,* 3rd edition, edited by J. Silverman, M.A. Suckow, and S. Murthy, 2014, CRC Press/Taylor & Francis.
- *Occupational Health and Safety in the Care and Use of Research Animals,* by the Committee on Occupational Safety and Health in Research Animal Facilities, 1997, National Academy Press.
- *Handbook of Laboratory Animal Science: Essential Principles and Practices,* 3rd edition, edited by J. Hau and S.J. Schapiro, 2011, CRC Press/Taylor & Francis.
- *Anesthesia and Analgesia in Laboratory Animals,* 2nd edition, edited by R. Fish, P. Danneman, M. Brown, and A. Karas, 2008, Academic Press.
- *Laboratory Mouse and Laboratory Rat Procedural Techniques: Laboratory Mouse Procedural Techniques* (Manual and DVD),

by J.J. Bogdanske, S. Hubbard-Van Stelle, M. Rankin-Riley, and B.M. Schffman, 2010, CRC Press.

- *Laboratory Animal Anesthesia*, 4th edition, by P.A. Flecknell, 2015, Academic Press.
- *Systematic Approach to Evaluation of Mouse Mutations*, edited by J.P. Sundberg and D. Boggess, 1999, CRC Press.
- *Mouse Phenotypes: A Handbook of Mutation Analysis*, by V.E. Papaioannou and R.R. Behringer, 2005, Cold Spring Harbor Laboratory Press.
- *Genetically Engineered Mice Handbook*, edited by J.P. Sundberg and T. Ichiki, 2006, CRC Press.

Guideline Documents

- *Guide for the Care and Use of Laboratory Animals*, 2011, National Academies Press.
- *Biosafety in Microbiological and Biomedical Laboratories*, 6th edition, HHS Pub. No. (NIH) 93-8395, 2020, U.S. Centers for Disease Control and National Institutes of Health (CDC/NIH).
- *Guide to the Care and Use of Experimental Animals: Canadian Council on Animal Care, Volume 1*, 2nd edition, 1993, Canadian Council on Animal Care.
- Guidelines for the veterinary care of laboratory animals: report of the FELASA/ECLAM/ESLAV Joint Working Group on Veterinary Care, H.-M. Voipio, P. Baneux, I. A. Gomez de Segura, J. Hau, and S. Wolfensohn, *Lab. Anim.*, 42, 1, 2008.

Periodicals

- *Comparative Medicine*. Published by AALAS.
- *Journal of the American Association for Laboratory Animal Science*. Published by AALAS.
- *Laboratory Animals*. Published on behalf of Laboratory Animals Ltd (www.lal.org.uk/) by Sage.
- *Lab Animal*. Published by Nature Publishing Group (www.nature.com/laban/).
- *ILAR Journal*. Published by the Institute for Laboratory Animal Research (academic.oup.com/ilarjournal).

electronic resources

- **Mouse Genome Database**: Supported by the U.S. National Institutes of Health and maintained through the Mouse Genome Informatics website of the Jackson Laboratory (www .informatics.jax.org), the Mouse Genome Database (MGD) is an integrated data resource for mouse genetic, genomic, and biological information. MGD includes a variety of data ranging from gene characterization and genomic structures, characteristics of inbred strains, and descriptions of mutant phenotypes to orthologous relationships between mouse genes and those of other mammalian species, and other related information.

- **The Mouse Brain Library (MBL)**: The MBL (www.mbl.org/) consists of high-resolution images and databases of brains from many genetically characterized strains of mice. The overall mission of the MBL is to systematically map and characterize genes that modulate the architecture of the mammalian central nervous system (CNS).

- **The Allen Brain Atlas**: The Allen Brain Atlas (https://mouse .brain-map.org/static/atlas) is an online, full-color, high-resolution anatomic reference atlas that is accompanied by a systematic, hierarchically organized taxonomy of mouse brain structures. A brain map, a mouse spinal cord atlas, a mouse brain connectivity atlas, and an atlas of the developing mouse brain are all available at the site as well (https:// portal.brain-map.org/).

- **International Mouse Phenotyping Consortium (IPMC)**: The IPMC (www.mousephenotype.org/) is an international effort by research institutions to identify the function of every protein-coding gene in the mouse genome in an effort to create a catalog of mammalian gene function that is freely available for researchers. The consortium is achieving this by systematically knocking out each gene in the genome and then conducting phenotyping tests across a range of biological systems.

- **International Mouse Strain Resource (IMSR)**: The IMSR is a searchable online database (www.findmice.org/) of mouse strains and stocks available worldwide, including inbred, mutant, and genetically engineered mice. The goal of the IMSR is to assist the international scientific community in locating and obtaining mouse resources for research.

- **Pathbase**: Pathbase (http://pathbase.net/) is a database of histopathology photomicrographs and macroscopic images derived from mutant or genetically manipulated mice. An international consortium of pathologists and veterinarians contribute to Pathbase. Images can be retrieved from the database by searching for specific lesion or class of lesion, by genetic locus, or by a wide set of parameters.

- **AALAS Community Exchange (ACE)**: Available through AALAS (www.aalas.org), ACE is a series of communities centered around various interest groups (e.g., animal researchers, Association of Primate Veterinarians, comparative medicine). Members post questions, have online discussions, and post resources on topics relevant to the community.

- **IACUC Central (formerly IACUC.org)**: Available through AALAS, IACUC Central (www.aalas.org/iacuc) is a resource for institutional animal care and use committee members and staff. Updated quarterly, IACUC Central organizes information into pages containing links to government agencies, databases, examples of institutional websites, training resources, and more.

- **AALAS Learning Library**: The AALAS Learning Library (www.aalas.org/education/aalas-learning-library) provides training that is essential for technicians, veterinarians, managers, IACUC members, and investigators working with animals in a research or education setting. The courses emphasize the appropriate handling, care, and use of animals, including rabbits.

- **Mouse Phenome Database:** The Mouse Phenome Database (phenome.jax.org) stores and organizes contributed datasets with detailed protocols on how the data were collected and provides tools for analysis. Data are from around the world and include phenotypic and genomic information from a variety of mice, including inbred strains, mutants, and others.

appendix

suggested cassette numbering system and some trimming suggestions

Cassette Number	Tissues	Trimming Suggestions
1	Heart	Hemisect (cut in half, longitudinally) to expose all the chambers and valves
	Thymus	Both lobes usually can be included intact
	Tongue	Cross section or longitudinal section, flat side down
	Diaphragm	Short strips can be sectioned on the edge
	Sternum	Place intact, deep, or internal side down in cassette, to section to marrow easily; trim off bony parts of ribs to facilitate sectioning
2	Lung, entire	Dorsal side down
	Trachea	Cross section at thyroid, or include intact for longitudinal section
	Esophagus	Include intact for longitudinal section
	Thyroid, parathyroid	Transect trachea at the level of thyroid or trim to evaluate lesions
	Aorta	Usually see longitudinal sections
3	Kidneys	Usually two sagittal sections of the left kidney plus two cross sections of the right kidney
	Adrenal glands	Small or important adrenals can be preserved by special cassettes (with smaller holes), sponges, or tea bags
	Lymph nodes	Frequently included with pararenal fat and adrenals

(Continued)

Cassette Number	Tissues	Trimming Suggestions
4	Salivary glands, exorbital lacrimal glands, auditory sebaceous glands, lymph nodes, mammary glands	Remove all these tissues in toto and place in cassette
5	Pancreas, lymph nodes, fat, vasculature	Strip fat, pancreas, mesentery, and lymph nodes from the GI tract to include in this cassette
6	Stomach	Section to include forestomach and glandular stomach
	Small intestine, cecum, colon	Include cross sections or segments according to needs or preferences
6 a, b, c	Swiss roll, open	The entire intestine can be opened, examined, fixed, and rolled into 1 or 2 deep cassettes
6 a, b, c	Swiss roll, closed	Alternatively, intact (closed) small intestine can be rolled into the first cassette; large intestine can be rolled into the second cassette; cecum and stomach can be sectioned into the third cassette
7	Liver, gall bladder	Section through the left lateral lobe, hilus to periphery; section through median lobe to include gallbladder
	Spleen	A small spleen may be included intact; a larger spleen can be hemisected along its long axis, and one or both halves evaluated
8 Female	Uterus, ovaries, vagina	Reproductive tract can be fixed flat and intact on a piece of paper. A small reproductive tract can be included intact in the cassette after fixation
	Urinary bladder	Often included in sections when the entire tract is included in the cassette
8 Male	Testes, epididymis, seminal vesicle + coagulating glands	Reproductive tract can be fixed flat and intact on a piece of paper. A small reproductive tract can be included intact in the cassette after fixation
	Prostate	Often included in sections when the entire tract is included in the cassette, or lobes can be dissected and fixed separately
	Urinary bladder	Often included in sections when the entire tract is included in the cassette or can be included separately
9	Skin	Cut 4-mm diameter ribbons of flat, fixed skin, parallel to hair growth and to long axis of mouse, to include clitoral or preputial glands (bilobed sebaceous glands in inguinal subcutis, near genital orifices)

(*Continued*)

Cassette Number	Tissues	Trimming Suggestions
	Mammary glands	Usually included with female skin sections, or mammary pads can be harvested and evaluated specifically
	+/− Decalcified leg	May be included in this cassette also. Trim tissue from the medial aspect so that the femur can be seen on the flat (cut) surface
10	Decalcified head	Section systematically using similar anatomic landmarks and consistent orientation. External ear canal openings and the eyes are useful landmarks. Holding the nose in one hand, using clean, single strokes, make 4 or 5 sections, from caudal to rostral, that usually fit in one cassette 1. Cut just caudal to ear canal for a section that includes cerebellum, medulla 2. Cut just rostral to ear canal for section to include middle ear and/or internal ear, pituitary, thalamus, hippocampus 3. Cut just caudal to eyes for section with cerebrum, usually hippocampus, thalamus 4. Cut just rostral to eyes for section with eyes, Harderian glands, oral cavity, molars 5. Nose section should be included also
11	Decalcified spine	Cervical–thoracic and lumbosacral spine segments (with muscle, vertebrae, spinal cord) usually can be accommodated in a total of two cassettes. Cross sections can be cut at anterior (or both) ends of each segment. Tissue can be cut cleanly from one side of the intervening segment to the level of vertebral bone to provide a flat surface on the paraffin block. This facilitates sectioning into deeper tissues, including the spinal canal and cord
12	Lesions	Ideally, lesions are included in cassettes 1–11 in their tissues of origin. Large lesions may require additional cassettes, which may be numbered according to the tissue of origin (e.g., 4a, 4b, for a large tumor in or near the salivary glands) or may be given a new number (e.g., 12). Lesions should be trimmed to include adjacent normal tissue for perspective and context and also to reflect gross measurements and photographs in order to improve correlation to gross findings

index

A

AAALAC, International, 70–71
AALAS, *see* American Association for Laboratory Animal Science
AALAS Community Exchange (ACE), 210, 216
AALAS Learning Library, 216
Abdominal palpation, 81
Abnormal gait, 96
ACE, *see* AALAS Community Exchange
Acidification (of water), 47
Acidophilic macrophage pneumonia, 97
ACLAM, *see* American College of Laboratory Animal Medicine
Adenosine triphosphate (ATP), 44
Adenoviruses, 107
Adrenal medullary tumors, 87
Aggressive behavior, 17
ALAT, *see* Assistant Laboratory Animal Technician
Alimentary system
  non-neoplastic conditions, 84–85
  tumors, 85
Allele, 48
Allen Brain Atlas, 215
Alopecia, 89
American Association for Laboratory Animal Science (AALAS), 209–210
American College of Laboratory Animal Medicine (ACLAM), 210–211
American Society of Laboratory Animal Practitioners (ASLAP), 210
American Veterinary Medical Association (AVMA), 189

Amyloidosis, 98–99
Analgesia
  administration, 186–187
  dosage, 187
  signs of pain, 185–186
Anemia, 88
Anesthesia
  compounds, 176
  methods and drugs, 176–177
    inhalant anesthetics, 177–178
    injectable anesthetics, 178–182
    special techniques, 182
  periprocedural care, 182–185
  post-procedural, 176
Angiectasis, 87
Anophthalmia, 96
Arthritis, 93
Arthropods, 117
ASLAP, *see* American Society of Laboratory Animal Practitioners
Aspiculuris tetraptera, 119–121
Assistant Laboratory Animal Technician (ALAT), 210
Athymic, 13
Autoclaving, 46
Automatic watering systems, 43, 47
AVMA, *see* American Veterinary Medical Association

B

Barbering, 17, 18
Barbiturate anesthetics, 178–182
Barrier, 33–36
Bedding

aspen wood chips, 32
cellulose, 31
chips, 31
contact, 30
contaminants, 31
cotton-based cellulose, 31–32
cotton fibers, 31–32
ground corncob, 31
noncontact, 30
pelleted, 31
recycled, 31
shredded, 31
specifications, 31
types, 31–33
ultra-high-absorbency, 32
wood shavings, 31
Behavior, 17–18
Beige mice, 14
Biocontainment barriers, 33
Biohazardous agents, 74
Biological materials
    human viruses, 150
    risks for humans and animals,
        148–150
    rodent pathogens, 149
Biosecurity barriers, 33
Bisphenol A (BPA), 26
Bordetella bronchiseptica, 110
Breeding, 47–48
    genetic monitoring, 54–55
    genetic terms, 48–49
    pheromones, 52–53
    records, 58
    schemes, 49–52
    timed pregnancy, 53–54

C

Cage cards, 55
Cages
    plastics, 26
    size recommendations, 27
Calorie-restricted mice, 32
Cannibalism, 19
Carcinoma, 93
Cardiac thrombi, 85
Cardiovascular system
    non-neoplastic conditions, 85–86
    tumors, 87
Cassette numbering system, 217–219
Cataracts, 95
Census records, 57
Cerebrospinal fluid (CSF), 96
Chattering, 80
Chemical sterilants, 43

Chlorination, 47
Chronic urinary obstruction, 98
Citrobacter rodentium, 111–112
Clinical medicine
    alimentary system
        non-neoplastic conditions, 84–85
        tumors, 85
    cardiovascular system
        non-neoplastic conditions, 85–86
        tumors, 87
    endocrine system
        non-neoplastic conditions, 87
        tumors, 87–88
    endpoints, 125–127
    hematopoietic and immune system
        non-neoplastic conditions, 88–89
        tumors, 89
    infectious diseases, 100–101 (*see also*
        Infectious diseases)
    integumentary system
        non-neoplastic conditions, 89–91
        tumors of skin and adnexae,
            91–93
    musculoskeletal system
        non-neoplastic conditions, 93–94
        tumors, 94
    nervous system
        non-neoplastic conditions, 94–96
        tumors, 97
    noninfectious diseases, 82–83
    nutritional status, 99–100
    physical examination, 80–82
    research
        aged mice, 128
        cancer, 127–128
        degenerative diseases, 128
        shock, 128
    respiratory system, 97
    spontaneous disease, 82–83
    stress-related changes, 100
    systemic/multisystem conditions
        amyloidosis, 98–99
        hyalinosis, 99
    urogenital system
        non-neoplastic conditions, 97–98
        tumors, 98
    veterinary supplies, 79–80
Clitoral/preputial gland neoplasms, 92
Clostridium piliforme, 112
CMAR, *see* Certification program for
        managers of animal resource
        programs
Coisogenic strains, 2
Colony management records, 57
Compassion fatigue, 75–76

Compound administration
    implantable cannulas and pumps,
        174–175
    intradermal (ID), 172
    intramuscular (IM), 170–171
    intraperitoneal (IP), 171–172
    intravenous (IV), 172–173
    oral (PO), 169–170
    retro-orbital (RO), 173–174
    subcutaneous (SC), 172
Congenic mice, 3, 5
Containment, 33–36
Corneal opacities, 95
Corynebacterium bovis, 112
Cryopreservation, 148

D

DEA, *see* Drug Enforcement
        Administration
Deer mouse, 15
Depopulation, 131
Dermatophyte test medium (DTM), 103
Diestrus, 168
Diets, 37–38
Disposable cages, 27
Drug Enforcement Administration
        (DEA), 190
DTM, *see* Dermatophyte test medium
Dwarf tapeworm, 75
Dyspnea, 97
Dystocia, 124–125

E

Ear punch, 56
Ear tags, 56
Ectromelia virus, 108–109
ELISA, *see* Enzyme-linked
        immunoabsorbent assay
Embryo transfer, 52, 148
Endocrine system
    non-neoplastic conditions, 87
    tumors, 87–88
Endometrial hyperplasia, 98
Enrichment, 20–21
Environment
    humidity, 38–39
    illumination, 39–40
    noise, 40–41
    temperature, 38–39
    ventilation, 39
    vibration, 40–41
Eosinophilic crystalline pneumonia, 97
Eosinophils, 88

Epicardial mineralization, 86
Epizootics, 101
Esophageal dilatation, 85
Estrus, 168
Euthanasia
    argon, 192–193
    barbiturate overdose, 191
    carbon dioxide, 191
    carbon monoxide, 191
    cervical dislocation, 192
    decapitation, 192
    ether, 192
    experimental endpoints, 190–191
    inhalant anesthetic overdose, 191
    laws/regulations/guidelines, 189
    management considerations,
        189–190
    microwave irradiation, 191
    nitrogen, 192–193
    potassium chloride, 192
    recommendations, 188
    scientific considerations, 190
Exogenous mammary tumor virus, 93
Experimental methodology
    cage transfer, 157–158
    restraint devices, 159
    restraint for manipulation, 158–159
Extraorbital gland, 9

F

FBR, *see* Foundation for Biomedical
        Research
Fecal flotation, 103
Fibroplasia, 85
Fight wound, 92
Filobacterium rodentium, 111
Filter tops, 27
Food and Drug Administration (FDA),
        70
Foundation for Biomedical Research
        (FBR), 212
Fur marking, 56
Fur mites, 117, 118

G

Gastrointestinal system, 7
Gestation, 19

H

Hantavirus, 74–75, 108
Harderian gland, 9
    neoplasms, 93

Health surveillance and monitoring
    diagnostic testing, 153
    principles, 150–151
    selection of test subjects, 151–153
Hemangioma, 87
Hemangiosarcoma, 87
Hematology, 12
Hematopoietic and immune system
    non-neoplastic conditions, 88–89
    tumors, 89
Hematopoietic neoplasms, 89, 98
Hepatocellular adenoma, 85
Herpesviruses, 108
Housing
    bedding, 30–33 (*see also* Bedding)
    caging, 25–28
    filter-top, IVC, 28–30
    filter-top, static racks, 28
    flexible-film/semirigid isolator, 30
Husbandry, 25
Hyalinosis, 97, 99
Hybrid mice, 2, 48
Hydrocephalus, 95
Hydrometra, 97
Hydronephrosis, 98
Hydrophobic sand, 166
Hypothermia, 182
Hypothyroidism, 88
Hysterectomy derivation, 147–148

I

Identification
    cage cards, 55
    complete and correct nomenclature,
        56
    ear punch, 56
    ear tags, 56
    fur marking, 56
    sex, 56
tattoo, 56–57
ILAR, *see* Institute for Laboratory
        Animal Research
Immunodeficient mice, 13–15, 37–38
IMSR, *see* International Mouse Strain
        Resource
Inbred mice, 2
Inbred strains, 48
Incisor dysplasia, 84
Individual health records, 57
Induced mutations, 2
Infectious diseases
    animals, 130
    approaches to elimination, 131
    bacterial agents

        Bordetella species, 110
        Citrobacter rodentium, 111–112
        Clostridium piliforme, 112
        Corynebacterium bovis, 112
        Filobacterium rodentium, 111
        Helicobacter species, 112–113
        mycoplasmas, 113
        Pseudomonas aeruginosa, 114
        Rodentibacter pneumotropicus,
            113–114
        Staphylococcus, 114–115
    burnout, 132
    depopulation, 131
    diagnostic methods
        detection of parasites, 103
        microbial culture, 103
        optimal testing, 102
        pathology, 102
        PCR, 102–103
        serologic methods, 102
    facility decontamination, 133
    fungal agents
        Pneumocystis murina, 115
    helminth, 119
    materials, 131
    medical treatment, 132–133
    nematodes, 119–121
    parasitic agents
        arthropods, 117
        Demodex musculi, 118–119
        fur mites, 117
        intestinal protozoa, 115–117
    prevention of spread, 129–130
    test and cull, 131–132
    viral agents, 103–104
        adenoviruses, 107
        ectromelia virus, 108–109
        hantaviruses, 108
        herpesviruses, 108
        LCMV, 108
        Mouse hepatitis virus (MHV),
            105–106
        Mouse rotavirus (MRV), 106–107
        Murine astrovirus (MuAstV), 107
        Murine norovirus (MNV), 104
        papovaviruses, 109
        parvoviruses, 104–105
        Theiler's murine encephalomyelitis
            virus (TMEV), 106
Inhalant anesthetics, 177–178
Injectable anesthetics, 178–182
Institute for Laboratory Animal
        Research (ILAR), 211
Institutional Animal Care and Use
        Committee (IACUC)

composition, 71
responsibilities, 72
Integumentary system
   non-neoplastic conditions, 89–91
   tumors of skin and adnexae, 91–93
International Mouse Phenotyping
   Consortium (IPMC), 215
International Mouse Strain Resource
   (IMSR), 215–216
Intestinal protozoa, 115–117
Intradermal (ID) injection, 172
Intramuscular (IM) injection, 170–171
Intraorbital gland, 9
Intraperitoneal (IP) injection, 171–172
Intravenous (IV) injection, 172–173
IPMC, *see* International Mouse
   Phenotyping Consortium
Isoflurane, 182
Isolators, 34
IVC system cage, 29

J

JAALAS, *see Journal of the American
   Association for Laboratory
   Animal Science*
*Journal of the American Association for
   Laboratory Animal Science*
   (JAALAS), 209
Juvenile mice, 18

K

Ketamine hydrochloride, 180
Klossiella muris, 12
Knockout/targeted mutation mice, 3
K virus, 109

L

Laboratory Animal Management
   Association (LAMA), 210
Laboratory Animal Science Professional
   (LAS Pro), 209
Laboratory Animal Technician (LAT),
   210
Laboratory Animal Welfare Training
   Exchange (LAWTE), 211
Lactate dehydrogenase–elevating virus
   (LDEV), 108
LAMA, *see* Laboratory Animal
   Management Association
LAS Pro, *see Laboratory Animal Science
   Professional*
LAT, *see* Laboratory animal technician

LAWTE, *see* Laboratory Animal Welfare
   Training Exchange
LCMV, *see* Lymphocytic
   choriomeningitis virus
Lee–Boot effect, 53
Leiomyoma, 94
Leiomyosarcoma, 94
Lymphadenomegaly, 88
Lymphocytic choriomeningitis virus
   (LCMV), 75, 108

M

Major histocompatibility complex (MHC),
   14
Major urinary proteins (MUP), 73
Malocclusion, 45, 84
Mammary adenoma, 93
Maternal behavior, 19
MBL, *see* Mouse Brain Library
Megaesophagus, 85
Metestrus, 168
MHV, *see* Mouse hepatitis virus
Microisolator cage, 27, 34
   vs. individually ventilated, 41
Minute mice virus (MMV), 104
MMV, *see* Minute mice virus
Mouse
   external features, 6
   gastrointestinal system, 7
   glands, eye, 9
   hearing and vocalization, 6
   respiratory system, 8
   skeleton, 6
   spleens, 9
   urogenital system, 7–8
   visual system, 6–7
Mouse Brain Library (MBL), 215
Mouse Genome Database, 215
Mouse hepatitis virus (MHV), 105–106
Mouse kidney parvovirus (MKPV), 105
Mouse papillomavirus, 109
Mouse parvovirus (MPV), 104
Mouse Phenome Database, 216
Mouse polyomavirus, 109
Mousepox, 108–109
Mouse rotavirus (MRV), 106–107
Mouse stereotypies, 19
Mouse urologic syndrome (MUS), 98
MPV, *see* Mouse parvovirus
MRV, *see* Mouse rotavirus
MuAstV, *see* Murine astrovirus
Mucometra, 97
Murine astrovirus (MuAstV), 107
Murine chapparvovirus (MuCPV), 105

Murine noroviruses (MNV), 104
Muscular dystrophy, 93
Musculoskeletal system
  non-neoplastic conditions, 93–94
  tumors, 94
Mus musculus, 15
Mus spretus, 15–17
Mycoplasma pulmonis, 111, 113
Myobia musculi, 117, 118
Myoepitheliomas, 85, 93, 109

N

NABR, *see* National Association for
  Biomedical Research
National Association for Biomedical
  Research (NABR), 212
Necropsy
  anatomic pathology, 193
  dissection and specimen collection,
    196–199
  equipment and materials
    camera, 195
    cutting board, 194
    decalcifying, 195
    eye protection, 194
    face protection, 194
    fixative, 195
    forceps, 194
    gloves, 194
    metric ruler, 194
    paper towels, 194
    protective uniform, 194
    scale/balance, 194
    scalpel blades, 195
    scissors, 194–195
    syringe and needle, 195
    workstation, 193
  external examination, 196
  histopathology, 200
  procedure, 195–196
  reporting and archiving data, 200
  trimming tissues for histology, 199–200
Neonatal mortality, 51
Neonates, 182
Neoplasms, 83
Nervous system
  non-neoplastic conditions, 94–96
  tumors, 97
Newborn fostering, 147
Nocturnal, 17
NOD, *see* Nonobese diabetic
NOD-Rag mouse (NRG), 15
NOD-SCID-gamma, 14
NOD.SCID mice, 14

Nomenclature, 3–5
Noninfectious diseases, 82–83
Nonobese diabetic (NOD), 14
Normative values
  biologic parameters, 9
  clinical chemistry, 9–11
  urinalysis, 11–12
Nude mice, 13
Nutrition, 44–47

O

Obese (ob) mutation, 5
Obesity, 83
Occupational health programs, 72–75
  allergy, 73
  experimental biohazards, 74
  hantavirus, 74–75
  noise, 73
  puncture and bite wounds, 73
Oily hair effect (OHE), 110
Orogastric gavage, 170
Osteosarcomas, 94
Outbred stock, 49

P

Pallor, 88, 89
Palpation, 81
Papovaviruses, 109
Paramyxoviruses, 110
Paresis, 96
Parturition, 19
Parvoviruses, 104–105
Pathbase, 216
Pedigree charts, 57–58
Periarterial inflammation, 85
Periodontal inflammation, 84
Peripheral lymphadenomegaly, 89, 90
Peromyscus species, 15
Personal protective equipment (PPE),
  36–37
Pharyngeal pouches, 87
Pneumonia virus of mice (PVM), 110
Polytropic strains, 105
PPE, *see* Personal protective equipment
Preputial fight wounds, 92
Preventive medicine
  cryopreservation, 148
  handling of animals, 143
  health report, 141–144
  quarantine
    acute use, 146
    practices, 145–146
    principles, 144–145

rederivation program, 146
    embryo transfer, 148
    hysterectomy derivation,
        147–148
    newborn fostering, 147
Proestrus, 168
Professional organizations
    AAALAC International, 211
    AALAS, 209–210
    ACLAM, 210–211
    ASLAP, 210
    FBR, 212
    ILAR, 211
    LAMA, 210
    LAWTE, 211
    NABR, 212
Pseudomonas aeruginosa, 114
Pseudopregnant mice, 52
Public Health Service (PHS), 69, 70
Pyometra, 97

R

Rack system cage, 29
Rag knockout, 105
Recombinant inbred mice, 2
Records
    breeding records, 58
    pedigree charts, 57–58
Regulatory agencies, 69–70
Reovirus 3 (REO3), 110
Respiratory system, 8, 97
Retro-orbital (RO), 173–174
Reverse osmosis, 47
Rhabdomyoma, 94
Rhabdomyosarcomas, 94

S

Sampling methods
    blood, 159–160
        cardiac puncture, 165
        decapitation, 165
        distal tail transection, 162
        facial vein, 160
        lateral tail vein, 160–162
        non-terminal blood collection
            techniques, 165
        retro-orbital sinus puncture,
            164–165
        saphenous vein, 162–164
    DNA analysis, 167–168
    feces, 167
    urine, 165–167
    vaginal swabs, 168–169

Sanitation
    cage cleaning, 41–43
    disease prevention, 53–54
    pest control, 44
    quality control, 43–44
    research equipment cleaning, 43
    room cleaning, 43
    water, 47
Scaly skin disease, 112
SCID, *see* Severe combined
        immunodeficiency
Sedation, 170
Semirigid isolators, 36
Serologic methods, 102
Serum amyloid A (SAA), 99
Severe combined immunodeficiency
        (SCID), 14
Sick mice, treatment of
    drug dosages, 122
    dystocia, 124–125
    open skin lesions, 122
    provision of food, 123–124
    supplemental fluids, 123
    supplemental heat, 124
    weak mice, 122–123
Specific pathogen–free (SPF), 148
Splenomegaly, 88
Staphylococcus aureus, 114–115
Strains, 2
Subcutaneous (SC) injection, 172
Surgical anesthesia, 180
*Syphacia obvelata vs. Aspiculuris
        tetraptera*, 121

T

Tattoo, 56–57
Theiler's mouse encephalomyelitis virus
        (TMEV), 106
Thymic atrophy, 100
Thymic lymphoma, 89, 90
TMEV, *see* Theiler's mouse
        encephalomyelitis virus
Transgenic mice, 2–3
Transplacental infection, 105–106
Transportation
    feed/water/bedding, 59
    health records, 59
    between institutions, 58–59
    within institutions, 60
    requirements during
        shipment, 59
    shipping container, 58–59
Tribromoethanol, 181
Tyzzer's disease, 112

U

Ulcerative dermatitis, 90, 91
Ultraviolet irradiation, 47
Urinalysis, 11–12
Urogenital system, 7–8
    non-neoplastic conditions, 97–98
    tumors, 98
US Department of Agriculture (USDA),
    59, 69

V

Vibrissae, 90
Visual system, 6–7

W

Weanling mice, 26
Well-being, 19–20
Western Mediterranean mouse, 15
Wet mount, 103
White noise, 40–41
Whitten effect, 53, 54
Wild-type, 48
Wire-bottom cages, 26

Z

Zoonotic diseases, 72–75